MATHEMATICS FRAMEWORK

for California Public Schools

Kindergarten Through Grade Twelve

Publishing Information

When the *Mathematics Framework for California Public Schools* was adopted by the California State Board of Education on November 8, 1991, the members of the Board were Joseph D. Carrabino, President; Marion McDowell, Vice-President; Kathryn Dronenburg, Paul Kim, Dorothy J. Lee, S. William Malkasian, Benjamin Montoya, Kenneth L. Peters, Joseph Stein, and Gerti B. Thomas. The framework was developed by the Mathematics Curriculum Framework and Criteria Committee and was recommended by the Curriculum Development and Supplemental Materials Commission to the State Board of Education for adoption. (See pages xii–xiv for a list of the names of the committee members and others who made significant contributions to this publication.)

The *Mathematics Framework* was prepared for photo-offset production by the staff of the Bureau of Publications, California Department of Education, under the direction of Theodore R. Smith, Editor in Chief, and in cooperation with Walter Denham, Manager, Office of Mathematics Education. The publication was edited by Edward O'Malley and was designed by Steve Yee, who also created the cover and prepared the layout with the assistance of Cheryl Shawver McDonald. Typesetting was provided by Carey Johnson, who was assisted by Jeannette Huff. Photographs were provided by Joe Donovan.

The framework was published by the California Department of Education, 721 Capitol Mall, Sacramento, California (mailing address: P.O. Box 944272, Sacramento, CA 94244-2720). It was printed by the Office of State Printing and distributed under the provisions of the Library Distribution Act and *Government Code* Section 11096.

© 1992 by the California Department of Education

ISBN 0-8011-1033-5

Ordering Information

Copies of the *Mathematics Framework for California Public Schools* are available for $6.75 each, plus sales tax for California residents, from the Bureau of Publications, Sales Unit, California Department of Education, P.O. Box 271, Sacramento, CA 95812-0271; FAX (916) 323-0823.

A list of mathematics publications available from the Department can be found on page 218. A complete list of Department publications can be obtained by writing to the address given above or by calling the Sales Unit at (916) 445-1260.

Contents

Foreword

Publication of the 1992 *Mathematics Framework for California Public Schools* answers a call from the President of the United States, the nation's governors, the National Council of Teachers of Mathematics, the Mathematical Sciences Education Board, and the teachers and mathematics educators who served on the Framework Committee. That call is to change what mathematics we teach, how we teach it, and to whom. The mathematics education that most Americans received in the past may have been good enough at that time but is unsatisfactory for today's students. The time is overdue for shifting the emphasis from an elaborate study of mechanics and procedure toward a deeper understanding of central ideas and a broader study of all the strands of mathematics.

Instructional materials and methods used in mathematics education must reflect the shift in content to avoid spoon-feeding students recipes for getting answers. Instead, students must become more responsible for formulating and solving problems and for thinking about and communicating important mathematical ideas. Teaching students to take more responsibility for mathematical thinking requires teachers to do more coaching and facilitating and less telling than most of us remember from our own days as students of mathematics. It also entails more demanding expectations for performance. Students will have to produce more than answers to prefabricated problems; they will have to use mathematics to investigate, analyze, and interpret realistic situations. And although assignments will be more demanding, they must also be more engaging and accessible to a wider variety of students than assignments given in traditional mathematics programs.

Progression through the mathematics curriculum is an unfulfilled promise for the majority of California's students. Gender, ethnicity, language, and poverty often diminish a student's chances to advance in mathematics. Although some individuals overcome the odds, all deserve a better chance. The curriculum, teaching, and assessment in mathematics education must be reconfigured so all students learn powerful mathematics each year in school.

The filtering effect of the traditional kindergarten through grade twelve mathematics program has produced an American public believing that only a few are good at mathematics and that it is satisfactory for most people to have negative attitudes toward mathematics (and the people who teach and practice mathematics). The next generations of the public are in our schools now. We must work with them to create a place for mathematics in our culture more like the place of music. That is, although only a few are very talented in music, most are good at it, and all can enjoy and participate in it. The beauty, fascination, and usefulness of mathematics make it accessible and important for every student.

The widely acclaimed 1985 *Framework* has laid a foundation for building the programs envisioned in this publication. Many teachers and schools have adopted better approaches for engaging more students in challenging mathematics activities that have cultivated more effective problem-solving and communication skills in the students. Now it is time to focus on harvesting the most important mathematical ideas from these fertile problem-solving experiences. This 1992 *Framework* calls for programs built from coherent units of study—units with situational and mathematical coherence. The priorities for breadth and depth of mathematics described in the pages that follow build on the *Curriculum and Evaluation Standards for School Mathematics*, published by the National Council of Teachers of Mathematics, which extends the philosophy of the 1985 *Framework*.

This *Framework* is devoted to change. It does not, however, describe programs that can be installed overnight. We must embark immediately, but we must be patient with ourselves and with each other. As better instructional materials and assessment tools become available, the programs envisioned will become practical for more teachers and students. Teachers will inevitably carry the greatest burdens. They deserve all the support that can be mustered, especially time with each other and with their colleagues who take leadership responsibilities for the reform of mathematics education. But times of change are also times of great excitement and opportunity. We stand our best chance of succeeding if we remember to enjoy the excitement, face our mistakes with good humor, and take pride in our pioneering accomplishments.

Joseph D. Carrabino (signature)

JOSEPH D. CARRABINO
President of the State Board of Education

MARION McDOWELL
Vice-President

KATHRYN DRONENBURG
PAUL KIM

DOROTHY J. LEE
S. WILLIAM MALKASIAN
BENJAMIN MONTOYA
KENNETH L. PETERS
JOSEPH STEIN
GERTI B. THOMAS

Bill Honig (signature)

BILL HONIG
State Superintendent of Public Instruction and Secretary and Executive Officer of the Board

Preface

The *Mathematics Framework* was developed between October, 1988, and October, 1991—a period in which the national mathematics education community reached an unprecedented degree of consensus. Both within and outside education, mathematics became known as the subject area that "has its act together." In addition, during that period critically important national publications began to appear, most notably the National Research Council's *Everybody Counts* (January, 1989); the *Curriculum and Evaluation Standards for School Mathematics* (March, 1989), issued by the National Council of Teachers of Mathematics (NCTM); and NCTM's *Professional Standards for Teaching Mathematics* (March, 1991). The resonance of the positions stated in those documents and by the Framework Committee was gratifyingly pronounced. As a result the *Framework's* development process was lengthened to take advantage of the national declarations, including some of the compelling language. This *Framework* incorporates the intentions for student learning already nationally proclaimed and adds specification to the structure and organization of instructional programs that teachers can use to meet the goal of mathematical power for all students.

There is overwhelming agreement that this *Framework* appropriately and accurately describes the mathematics programs that should be established in our schools. At the same time it is a framework for supporting change rather than a description of what can be installed overnight. The question is how best to proceed in every location in the educational system. The challenge is immense; yet, as this document is being published, there are opportunities to take concrete steps toward *Framework* implementation and full program reform.

Performance assessment is under accelerated development not only in California but across the country as well. Major foundation support is linking other states and California in a process of setting performance standards that will make our expectations for students outcomes clear for students and parents, teachers, and administrators. The developing new state assessment system will be built on these performance standards and assessments.

Equally important, several major curriculum development projects are under way at all grade levels. There is strong reason to believe that by the middle of this decade all schools will have new instructional materials developed from scratch, not as revisions of traditional programs that have for so long failed our students and teachers. In 1988 the California Department of Education initiated the development of a Math A course to replace ninth-grade general mathematics. Hundreds of high school teachers have used preliminary versions of this material, and hundreds more will be using the finished materials by 1993. In the summer of 1990, the Department began to sponsor elementary curriculum replacement workshops in which four- to six-week units developed by educators to meet the design goals of the NCTM's *Standards* and the *Framework* were used. The overwhelming demand for these units and additional units of the same quality is a strong signal that by 1995 the mathematics teachers in California's elementary schools and middle schools can expect the complete replacement of the mathematics curriculum. In 1992 early versions of high school courses designed to have virtually all students progress through a program that satisfies university admission requirements and does not filter out half of them are being pilot-tested in dozens of classrooms.

But the availability of performance standards, assessment tasks, and new curriculum units is not enough for complete program reform at a school. The critical factor in reform is support for teachers, especially in the form of extra sponsored time. Change takes time, and the typical faculty will need three to five years of sustained commitment and support to make the full transition to an empowering program. Federal and state agencies are now creating the necessary structures for committed teachers to make the changes successfully. These structures will provide cost-effective opportunities for schools and school districts to make the substantial investments in human capital that are required. The clearest opportunity in California lies in the middle grades, for which a Mathematics Renaissance Initiative involving 78 schools was launched in 1991. Each participating school started by releasing two mathematics teachers for 12 days during the academic year. These teachers participated in professional development activities led by accomplished veteran teacher-leaders. By the time this *Framework* appears in published form, every middle school in California will have received an invitation to participate in the initiative.

The time for small-scale experiments is past. We know what is required. As the *Framework* emphatically states, the pace of change will inevitably vary, and the components of change must move in concert for each school. Nevertheless, with clear vision and a united leadership, the opportunity is ours, and we urge every participant in the reform of mathematics education to respond in the fullest possible measure.

SALLY MENTOR
Deputy Superintendent
Curriculum and Instructional
Leadership Branch

WALTER DENHAM
Manager
Office of Mathematics Education

FRED TEMPES
Associate Superintendent
and Director
Curriculum, Instruction,
and Assessment Division

Acknowledgments

The *Framework's* development represents an unusually broad-scale effort. Many educators contributed to the document's original development. Over 500 individuals responded at length to the field review draft. In addition, several dozen teachers and professors subsequently gave generously of their time in making revisions and additions to the *Framework,* which was approved by the State Board of Education.

Overall guidance was provided by the Mathematics Chair for the Curriculum Commission:

Ann Carlyle, Ellwood Elementary School, Goleta Union Elementary School District

The Mathematics Curriculum Framework and Criteria Committee, which consisted of 15 eminent California mathematics educators and mathematicians, was chaired by:

Gerlena R. Clark, Curriculum Programs and Instructional Technology, Office of the Los Angeles County Superintendent of Schools

The other committee members were:

Robert Alpert, Vista Grande Elementary School, San Ramon Valley Unified School District

Dani Brutlag, Investigations Project, Office of the President, University of California

Ruth Cossey, School of Education, Stanford University

Philip Curtis, Department of Mathematics, University of California, Los Angeles

Jorge de la O, Glen City Bilingual Market School, Santa Paula Elementary School District

Theresa Hernandez-Heinz, Mission High School, San Francisco Unified School District

Robin Kato, Jose Ortega Elementary School, San Francisco Unified School District

Kenneth Koppelman, Montgomery High School, City of Santa Rosa High School District

Judith Kysh, Northern California Mathematics Project, University of California, Davis

Judith Mumme, Tri-County Mathematics Project, University of California, Santa Barbara

Antoinette (Annie) Podesto, Office of Curriculum and Staff Development, Stockton City Unified School District

Elaine Rosenfield, Del Mar Elementary School, San Luis Coastal Unified School District

Alan Schoenfeld, School of Education, University of California, Berkeley

Gary Tsuruda, Stanford Middle School, Palo Alto Unified School District

The final stages of the *Framework's* development were directed by:

Phil Daro, California Mathematics Project, Office of the President, University of California

with critical support from:

Dick Stanley, Professional Development Program, University of California, Berkeley

and the Chair of the Curriculum Commission:

Elizabeth Stage, California Science Project, Office of the President, University of California

Principal writers of the document approved by the State Board of Education were:

Meg Holmberg and **Tim Erickson**

Staff support from the California Department of Education was provided throughout by:

Joan Akers, Walter Denham, and **Tom Lester,** Office of Mathematics Education

Janet Chladek, Jerry Cummings, and **Glen Thomas,** Curriculum Framework and Textbook Development Unit

Members of the Mathematics Subject Matter Committee, Curriculum Development and Supplemental Materials Commission, who were responsible for overseeing the development of the *Framework*, including the field review, were:

Del Alberti, Washington Unified School District

Ann Carlyle, Ellwood Elementary School, Goleta Union Elementary School District

Ann Chlebicki, Instructional Services Office, Saddleback Valley Unified School District

Bruce Fisher, Fortuna Elementary School, Fortuna Union Elementary School District

Yvonne Johnson (Chair), Office of Instruction, Hayward Unified School District

Maria Lopez-Freeman, Montebello High School, Montebello Unified School District

Chiyomi Masuda, Albany Middle School, Albany City Unified School District

Vivian Lee Ward, Sequoia High School, Sequoia Union High School District

Sincere appreciation is extended to the following for their cooperation in obtaining photographs for the *Mathematics Framework:*

Faculty and students at Frontier Elementary School, Rio Linda Union Elementary School District; and San Juan High School, San Juan Unified School District, Citrus Heights

Grateful acknowledgment is made to the following publishers and organizations for permission to quote from copyrighted materials:

For selected quotations from *Curriculum and Evaluation Standards for School Mathematics.* Reston, Va.: The National Council for Teachers of Mathematics, Inc. Copyright 1989. Reprinted with permission from the National Council for Teachers of Mathematics. For selected quotations reprinted with permission from *Everybody Counts: A Report to the Nation on the Future of Mathematics Education.* © 1989 by the National Academy of Sciences. Published by National Academy Press, Washington, D.C. For selected quotations from Stigler, James W., and Harold W. Stevenson. "How Asian Teachers Polish Each Lesson to Perfection." Reprinted with permission from the Vol. 15, No. 1 (spring, 1991) issue of the *American Educator,* the quarterly journal of the American Federation of Teachers. For selected quotations from *Mathematics Assessment: Myths, Models, Good Questions, and Practical Suggestions.* Reston, Va.: The National Council of Teachers of Mathematics. Copyright 1991. Reprinted with permission from the National Council for Teachers of Mathematics. For selected material from Brutlag, Dani. *Beyond the Surface.* Oakland, Calif.: Office of the President, University of California. Copyright 1991. Reprinted with permission from Office of the President, University of California. For selected quotations from Resnick, Lauren B. *Education and Learning to Think.* Reprinted with permission from *Education and Learning to Think,* 1987. Published by National Academy Press, Washington, D.C. For selected quotations reprinted with permission from *Reshaping School Mathematics: A Philosophy and Framework for Curriculum,* 1990. Published by National Academy Press, Washington, D.C. For a selected quotation reprinted by permission of the publisher from Cohen, Elizabeth G., *Designing Groupwork: Strategies for the Heterogeneous Classroom* (New York: Teachers College Press, © 1986 by Teachers College, Columbia University. All rights reserved.), quote from p. 91. For a selected quotation from Resnick, Lauren B. "Defining, Assessing, and Teaching Number Sense," in *Establishing Foundations for Research on Number Sense and Related Topics: Report of a Conference.* Edited by Judith T. Sowder and Bonnie P. Schappelle. Copyright 1989. Reprinted with permission from the Department of Mathematics, San Diego State University. For a selected quotation from Dossey, John A., and others. The *Mathematics Report Card: Are We Measuring Up?* Princeton, N.J.: Educational Testing Service. Copyright 1988. Reprinted with permission from the Educational Testing Service. For a selected quotation from Russell, S. J., and others. *Beyond Drill and Practice: Expanding the Computer Mainstream.* Reston, Va.: Council for Exceptional Children. Copyright 1989. Reprinted with permission from the Council for Exceptional Children. For a selected quotation from Thurston, William P. "Mathematics Education." Reprinted from *Notices of the American Mathematical Society,* Vol. 37, No. 7, September, 1990, by permission of the American Mathematical Society. For a selected quotation from Wiggins, Grant. "Teaching to the (Authentic) Test." Reprinted from *Educational Leadership,* April, 1989. Copyright by ASCD (Association for Supervision and Curriculum Development). For selected quotations reprinted with permission from *On the Shoulders of Giants,* © 1990 by the National Academy of Sciences. Published by National Academy Press, Washington, D.C. For material selected from "Do Bees Build It Best?" © Interactive Mathematics Project, 1991 (San Francisco State University and the Lawrence Hall of Science, University of California at Berkeley).

I n t r o d u c t i o n

The 1985 edition of the *Mathematics Framework for California Public Schools, Kindergarten Through Grade Twelve* (California Department of Education), established the goal of mathematical power for all students: the ability to discern mathematical relationships, reason logically, and use mathematical techniques effectively. Since 1985 a number of other influential publications have affirmed and enhanced that goal, including the *Mathematics Model Curriculum Guide, Kindergarten Through Grade Eight* (California Department of Education, 1987); *Everybody Counts: A Report to the Nation on the Future of Mathematics Education* (National Academy Press, 1989); the *Statement on Competencies in Mathematics Expected of Entering Freshmen* (California Department of Education, 1989); the *Curriculum and Evaluation Standards for School Mathematics* (National Council of Teachers of Mathematics, 1989); *Reshaping School Mathematics* (National Academy Press, 1990); *A Call for Change: Recommendations for the Mathematical Preparation of Teachers of Mathematics* (Mathematical Association of America, 1990); and *Professional Standards for Teaching Mathematics* (National Council of Teachers of Mathematics, 1991). A number of multiyear projects for primary through graduate education have begun to address the challenges outlined in those documents.

As a result of national, state, and local initiatives, evidence of change in mathematics education is accumulating. In many schools throughout the country and in California, teachers are emphasizing problem solving, providing students with hands-on, highly interactive learning experiences. Their students are not limited to learning arithmetic and algebra but study all of the different strands of mathematics, including measurement, statistics, and logic. In addition, computers and calculators are becoming common classroom tools; and innovations in student assessment, such as open-ended questions and portfolios, are finding their way into classroom, district, and state assessment programs. Reform in mathematics education has direction, coherence, and momentum.

Reform is necessary. In current programs too few young people leave school mathematically powerful. To address this problem, this *Framework* asks that teachers and students raise their expectations and expand their vision of what can happen in a mathematics classroom and what can appear in mathematics instructional materials. *All* students are capable of the level of work described here, but time, hard work, and courage will be needed to make this vision a reality.

What's New in 1992?

The 1992 *Framework* reinforces the momentum toward reform. Building on its predecessor, this *Framework* elaborates the concepts and recommendations contained in the 1985 edition. It also extends those concepts into a comprehensive vision for mathematics education, one that serves the larger purposes of schooling: to equip students with the reasoning tools they need as good citizens; to prepare students for successful work lives; and to develop students' personal capacities to enjoy and appreciate mathematics.

Mathematical Power

Echoing the 1985 *Framework*, this document reasserts the goal of mathematical power for all students and emphasizes the phrase "for all students." Many of the recommendations

here are motivated by a concern for equity—giving every student in California fair access to mathematics education. Included are females and males; rich, poor, and middle class; descendants from all parts of the world; speakers of Mandarin, English, Arabic, Spanish, and the more than 200 other first languages of U.S. citizens.

In Chapter 1 mathematical power is described in this way: *Mathematically powerful students think and communicate, drawing on mathematical ideas and using mathematical tools and techniques.*

What do those words mean in this context?

- *Thinking* refers to intellectual activity and includes analyzing, classifying, planning, comparing, investigating, designing, inferring and deducing, making hypotheses and mathematical models, and testing and verifying them.[2]
- *Communication* refers to coherent expression of one's mathematical processes and results.
- *Ideas* refer to content: mathematical concepts such as addition, proportional relationships, geometry, counting, and limits.
- *Tools and techniques* extend from literal tools such as calculators and compasses and their effective use to figurative tools such as computational algorithms and making visual representations of data.

Mathematically powerful students use these four components, the dimensions of mathematical power, to do something meaningful. That is, *mathematically powerful work is purposeful.* This purpose need not be utilitarian; on the contrary, students may be motivated by curiosity or whimsy—as long as they have a sense of purpose.

Three additional expectations for students are that they work successfully both individually and with others; come to appreciate mathematics in history and society; and exhibit positive attitudes towards mathematics, working with confidence, persistence, and enthusiasm.

Mathematical Performance

What about performance? How well should students be expected to learn mathematics? Student work should demon-

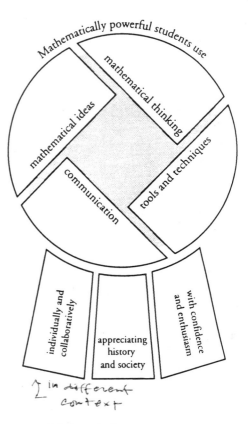

"Mathematical power, which involves the ability to discern mathematical relationships, reason logically, and use mathematical techniques effectively, must be the central concern of mathematics education and must be the context in which skills are developed."

—*Mathematics Framework,* 1985, p. 1[1]

---

[1]Complete citations for all quotations and other references appear in the Selected References.

[2]The *Curriculum and Evaluation Standards for School Mathematics* characterizes thinking in three of the first four standards: reasoning, problem solving, and making connections.

strate mathematical power in all its dimensions. This is a broad demand. The program must not waver in its insistence on good thinking, clear communication, proficiency with tools, and deep understanding of the mathematical content. Students must show that they can use mathematics to do something meaningful. At the same time the program must support students, giving them a realistic chance to show what they can do. Besides clear standards and high expectations, this support includes adequate preparation for sophisticated work and time to work on it, access to tools and colleagues, a developing tradition of draft and revision, and, finally, tasks that are worthy of the kind of work that demonstrates mathematical power.

Large Assignments

Creating assignments that encompass all the dimensions of mathematical power has two important implications. First, many assignments will have to be larger than those of the past, and students may need days or weeks to complete their work. Second, because the assignments will be large, students will have to work on the assignments outside class once the students are old enough. This document calls these larger projects *investigations*, a natural extension of the situational lessons called for in 1985.

However, exercises or smaller problems should not be eliminated, nor should students be expected to discover everything by themselves. Teachers and curriculum developers are expected to use their creativity and experience to provide students a balanced diet of different types of work: short exercises, interesting problems, collaborative tasks, and larger, long-lasting projects. The program will also benefit from varied instructional modes: teachers will enable and facilitate as well as demonstrate and instruct. Achieving a balance requires moving from exercises toward large projects and from direct instruction toward facilitating. Exercises will, therefore, be embedded in larger projects where possible, and instruction will more often help students learn specifically what they need to finish a large piece of work.

Complete Work

Traditionally, students have worked in only one dimension—with tools and techniques. Occasionally, students' thinking (how did she analyze this problem?) or ideas (how well does he understand sampling?) have been evaluated independent of right answers. But clear communication beyond format and legibility has almost never been a concern. Now, however, when a student successfully finishes a large-scale project that demonstrates all the dimensions of mathematical power, that accomplishment will be referred to as *complete* mathematical work. Grades might be based on complete work rather than on tasks (such as computational exercises and word problems) that, by their nature, do not demonstrate all of the dimensions.

The word *complete* is crucial because it implies not only the scope of the work but also its revisability. If a student's work does not meet quality standards, it needs revision; and students are expected to revise their work until it meets those standards. At the same time students have a right to assistance and encouragement as they improve their work. This policy shows respect for students, rewards persistence, and eliminates early failures that convince so many students that they are not suited to the study of mathematics.

Mathematical Content

What will the mathematical content of this complete work be? The 1985 *Framework* identified seven strands of mathematical content: *number, measurement, geometry, patterns and functions, statistics and probability, logic,* and *algebra.* The 1992 document endorses those strands and adds another, *discrete mathematics.*[3] In addition, this document expands *logic* to *logic and language,* acknowledging the importance of language in presenting mathematical ideas unambiguously; and changes *patterns and functions* to *functions,* recognizing that patterns are part of every strand. (Patterns are promoted to an even more important role in the curriculum and are discussed later under *unifying ideas.*)

Strands play a pivotal role in the design and implementation of complete mathematical programs. They represent continuous threads running throughout the curriculum, each

"The reader should keep in mind, however, that the division into strands is somewhat arbitrary, that more or fewer strands could be named, and that the strands frequently overlap."

—*Mathematics Framework,* 1985, p. 8

---

[3]This unfamiliar title includes things teachers at all grade levels have been doing for years. *Discrete* implies emphasis on separate (discrete) entities rather than on measures of continuous quantities—on questions of *how many* rather than *how much.* In third grade, for example, Boris can be asked in how many ways he can dress if he has three shirts and two pairs of pants (six); and in seventh grade students can be asked to design a tournament. This discrete mathematics includes topics such as combinatorial counting principles (how to count permutations and combinations for probability problems) and discrete structures (such as networks and trees). A resource for teachers is *Discrete Mathematics Across the Curriculum, K–12,* the NCTM yearbook for 1991.

being developed in appropriate ways at all grade levels, kindergarten through grade twelve. Strands help evaluate whether the mathematical content of a curriculum is broad enough and well balanced at all grade levels.

However, by themselves the strands are not sufficient to help identify the most important ideas for the curriculum. This *Framework* provides guidance here through another way to look at mathematical content—through *unifying ideas*, which focus understanding on a few very important and deep mathematical ideas and are central goals for student learning in each grade span.

For example, *proportional relationships* is a unifying idea for the middle grades. Elementary students work with proportions, as do high schoolers; but in the middle grades students focus most intently on proportional relationships as a mathematical principle that unifies a broad range of concepts and applications. Another unifying idea is *patterns*, which begins to play a role in the elementary grades and whose role extends in middle school to include generalizations of many kinds.

Unifying ideas allow a focus on mathematical themes that bridge many strands. The example of *proportional relationships* draws from virtually every strand: scale drawings from *measurement*; sampling from *statistics*; percent and ratios from *number*; linear relationships from *functions*; solution of proportions from *algebra*; and similarity from *geometry*. (See pages 122–25 in this *Framework*.) Unifying ideas bind the curriculum together through the year and give a focus to understanding.

In 1989 the National Council of Teachers of Mathematics (NCTM) adopted the *Curriculum and Evaluation Standards for School Mathematics*. These standards include a classification similar to the 1985 strands. In keeping with the national consensus, this *Framework* incorporates the standards as desired outcomes of the curriculum for kindergarten through grade twelve. The strands used in this *Framework* correspond closely to many of the individual NCTM standards, and the unifying ideas used in this *Framework* represent themes that appear in several different strands (and standards). Unifying ideas add a different dimension to content description and help to integrate subjects too often kept separate. These issues of content are discussed fully in Chapter 3.

"... Students should come to understand and appreciate mathematics as a coherent body of knowledge rather than [as] a vast, perhaps bewildering, collection of isolated facts and rules."

—*Curriculum and Evaluation Standards for School Mathematics*, p. 91

Curriculum Units

How will strands and unifying ideas appear in the curriculum? The curriculum will be divided into large chunks called *units*. Sizable units prevent the breaking up of large mathematical ideas into disconnected bits and make room for more opportunities for complete work. Units are about coherent blocks of subject matter that typically incorporate more than one unifying idea as well as several strands.

For example, a unit about *growth* bridges the unifying ideas of *proportional relationships* and *patterns and generalization*. Some units have subject matter that is relatively concrete and realistic, such as the *mathematics of growth*, while the subject matter of others is more abstract, such as *coordinate systems*. Some units support others by providing elements of mathematics that other units require.

While other countries have concept-sized lessons, often lasting weeks, the U.S. has lesson-sized concepts, lasting one day plus homework.

—Adapted from McKnight, *The Under-achieving Curriculum*, p. 89

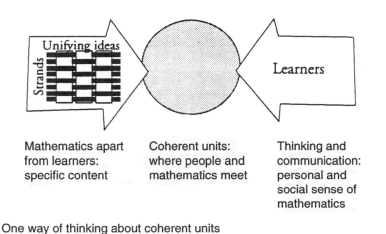

Mathematics apart from learners: specific content

Coherent units: where people and mathematics meet

Thinking and communication: personal and social sense of mathematics

One way of thinking about coherent units

Units will typically last from one to six weeks according to the subject matter and the age of the students. They will feature several opportunities (e.g., investigations) for students to produce complete mathematical work. In addition, students will work on smaller problems to obtain different perspectives on the ideas in the large projects and will do exercises for relevant practice. Thus, work in units is not limited to exploring and investigating; teachers and students will need to demonstrate straightforward mathematical tools, explain conventions and notation, and summarize. Most important, the unit is coherent because its content and assignments fit the subject matter, the direct instruction and demonstration support the large assignments, and the activities relate to one another through a common context or through well-designed summary lessons.

An appropriate example here might be an eighth grade unit on games of chance that focuses on the mathematics of *fairness*. The unit would have classroom activities about probability, of course, and problems and experiments for students to do—individually and in groups, both within and outside the classroom. There would probably be one or more larger organizing projects in the unit. For example, students might invent, create, and write intelligible rules for a game and analyze its fairness, using mathematical insight about chance. They would write about the mathematics behind their game and the *mathematical* reasons for choices they made in the design of the game.

In producing the game and its accompanying literature, students would demonstrate all of the dimensions of mathematical power: their thinking (design, conjecture, analysis); their use of ideas (proportions in game probabilities); their use of tools and techniques (computation, simulation); and their ability to communicate (rules and justification). The unit itself is based on a mathematical subject (*the mathematics of fairness*); it develops several unifying ideas (*proportional relationships* and *multiple representations*); and it interweaves material from several strands (*number, logic and language,* and *statistics and probability*).

Assessment of Mathematical Power

How can it be known whether student work demonstrates mathematical power? A traditional test is a snapshot of a student's ability to recall facts or procedures. By itself the test provides a limited picture of a student's mathematical power, usually focusing on his or her mastery of mathematical tools and techniques and memory of facts rather than on mathematical thinking, understanding, and skill in communicating.

More powerful assessment tools are needed to evaluate the work of students. Instructional materials aligned with this *Framework* and the *Curriculum and Evaluation Standards* should include appropriate assessment materials and examples of student work.

What is appropriate? Important advances have occurred in providing alternative means of assessing performance in mathematics. For example, the California Assessment Program (CAP)

8

has been changing from reliance on prestructured answers to performance assessment techniques, including open-ended items and portfolio assessment. Developers of national tests are moving in the same direction. Teachers throughout California are having students present and write more; are observing their students working in groups; and are learning to manage portfolios of student work. They are finding that they know more about student thinking and understanding than ever before.

Mathematics assessment in the 1990s will focus increasingly on assessing large pieces of work rather than smaller ones. Assessment and instruction will be integrated more effectively. Teachers can raise standards and expectations, asking thoughtful questions that allow for thoughtful (and unexpected) responses. In their turn students can become more involved in the assessment process; and teachers and students alike can learn to evaluate work holistically, using quality standards. Most important, teachers can learn to assess student work rather than the student and focus on what students know rather than on what they don't know. This shift of emphasis will free students from the stigmata of past failures and facilitate their taking responsibility for their own learning. All of these issues are elaborated in Chapter 2, pages 39–74.

The reader can learn more about what the California Assessment Program has done and see examples of student work in *A Sampler of Mathematics Assessment* (Sacramento: California Department of Education, 1991).

A Framework for Change

This *Framework* provides a road map for change and offers many paths that lead toward the destination. It offers guidance for a journey that began with previous frameworks. No shortcuts can be found here, no easy route to full and immediate attainment of goals. Rather, a multitude of possibilities appropriate to a wide range of circumstances present themselves. Some of these possibilities must be made real and must be tested in the classroom on students who represent the diversity of California's multicultural population.

This process of change has begun. In response to the *Curriculum and Evaluation Standards for School Mathematics*, new materials, with integrated assessment and new instructional techniques, are being developed and tested nationwide in classrooms populated by diverse students. Even so, the ideas in this *Framework* have not been implemented all together on a wide scale.

Therefore, genuine change in schools is necessarily uncertain and expected to be uneven. As new courses evolve, they will become accessible to more students, supporting full development of all students' mathematical potential: the especially talented, the traditionally disenfranchised, and students who are both or neither. Time, resources, and problem solving will be needed to reinvent the curriculum and thereby serve all students well.

The efficacy and acceptance of new programs will depend not only on the quality of the curriculum but also on a variety of other factors. Therefore, action needs to occur on many fronts. Elements essential to a good start are the following:

- Knowledge and commitment at the local level
- Availability of appropriate instructional materials and supplies
- External assessments consistent with instructional programs
- Cooperation among different parts of the educational system
- Teacher involvement in planning
- Professional development for teachers

The changes called for in this *Framework* cannot be accomplished by simply discarding the old in favor of whatever looks new or fashionable. The good that has been achieved must be conserved through the selective targeting of critical problems and efforts to ensure that the important mathematics will not be lost. Student work must embody all the components of mathematical power and, over time, must employ the full scope of the content laid out in chapters 3, 4, and 5.

Teachers and Change

With the concurrence of administrators and other mathematics educators not assigned to the classroom, the teachers involved in developing this *Framework* envisioned their role in implementing change as follows:

> Ultimately, it is the teachers who will make new mathematics programs a reality in the classroom. As teachers we have already begun to play our part in thousands of classrooms across the state. Many of us use manipulatives and cooperative learning; we assign complex problems and maintain portfolios; we think about the mathematical power of all our students; and we are changing the ways that we and our students think about mathematics.

Any new program makes demands on teachers. This one asks us to look at mathematics in a new way and to redefine our role in the classroom. To make the program work, we will need support. One of the most important areas of support is in time:

- Time to collaborate with peers: plan together, visit one another's classrooms, review student work, and make judgments about program strengths, weaknesses, and overall effectiveness
- Time to prepare for class and plan
- Time for continuing to learn mathematics
- Time to respond to student work
- Time to meet with teacher leaders and outside consultants, who can raise important questions and guide the inquiry process
- Time for school planning and organizational meetings
- Time to attend professional meetings and conferences away from school

The additional time provided for such professional responsibilities must be quality time. Time after school and unpaid Saturdays are not enough. Teachers will need at least ten days of extra professional time annually, preferably spread throughout the school year, to implement the programs described here and to share their successes and disappointments. And emerging site leaders will need even more time.

But teachers' needs go beyond time. Change of the scope presented here requires totally new instructional materials—the products that come from the industry, imagination, and flexibility of developers working in collaboration with practicing teachers. Students will need tools, too, ranging from blocks to computers, to use as they do their mathematical work.

New materials and equipment by themselves are not enough, however, to address the long-term needs of students. Teachers are asking questions such as the following: "How do I develop a cohesive plan for what my students experience?" "How do I address the needs of all my students in a diverse classroom?" "How do I know that my students are developing mathematical power and not just having fun?" No one can answer these questions alone; neither will mandating a workshop take care of the problems. Teachers have to work together in an environment of continual quality professional development.

"... American teachers are overworked. . . . The full realization of how little time American teachers have when they are not directly in charge of children became clear to us during a meeting in Beijing. . . . Beijing teachers teach no more than three hours a day, unless the teacher is a homeroom teacher, in which case the total is four hours. . . . The situation is similar in Japan. According to our estimate, Japanese elementary school teachers are in charge of classes only 60 percent of the time they are at school."

—Stigler and Stevenson, "How Asian Teachers Polish Each Lesson to Perfection," 45

Teachers may be expected to teach mathematics not yet invented when many teachers were in school, such as discrete mathematics; or mathematics that teachers may not have studied, such as statistics. The solution to these difficulties will be a combination of upgrading the mathematical understanding and learning of both teachers and students. Part of a balanced program of staff development will include mathematics content; but genuinely not knowing the answer and modeling good questioning and learning behavior gives students a clear message about process and about the value of thinking. The key is to recognize what must be learned and to share experiences with students.

Teachers also receive mixed messages. For example, although the *Framework* sets forth goals for mathematical power, local testing mandates may continue to emphasize narrow skill attainment. Teachers know implicitly that what is tested is what is important in the eyes of those in authority. Professional development may be directed toward implementing the *Framework*; but if teachers are held responsible for increasing scores on tests for low-level skills, what will receive emphasis? Once teachers and administrators agree on the goals of mathematical power and clarify the differences between those goals and the traditional ones, they can work together to match external testing to the system's real goals for its students.

The teachers who helped develop this *Framework* identified the most difficult task confronting teachers of mathematics as follows:

> The biggest challenge for us as teachers will be in how we practice our profession. It will take courage for us to abandon our traditional roles as gatekeepers and sources of information and approval and become the "chief questioners," the facilitators and catalysts for students. We must reflect, experiment, and accept uncertainty as part of the new professional norms. Certainly, sharing doubts and confusions with colleagues is difficult. We may be tempted to lose patience or abandon hope but ultimately will be empowered through this reform. Only through professional exploration and examination can we gain the experience and confidence necessary to implement the new programs and take the curriculum into our own hands.

Students and Change

Students are also members of the school community; they must change as well. Like their teachers, students will learn to work in different ways and to play different roles. And like their teachers, they will need "staff" development in these new ways of working and studying. Enlisting students, their parents, and the community in the reform process can make the outcome as well as the process better for all.

In the programs called for in this *Framework*, students will be asked to do more work and take more responsibility for organizing and revising their work outside class. Although students will be working harder and receiving more difficult assignments, the assignments will be more accessible than those typical of traditional programs. Parents and community members can make a major contribution by supporting community understanding of what is happening in education and finding ways of helping students—for example, by identifying places and times for students to work together outside class.

Over the next decade local, state, and national initiatives will likely impose higher and higher expectations on programs and students. Any new national assessment instruments will be based on the NCTM *Standards;* and California's efforts will be based on this *Framework,* which extends the national document. These new efforts will increasingly involve assessing student work on projects from regular instruction, portfolios, and open-ended responses rather than multiple-choice items and micro-tasks. To be ready, California students need to develop—and everyone needs to nurture—a mathematical work ethic that requires self-discipline and efforts to meet a high-quality standard.

Time Required for Change

This *Framework* states goals for California's mathematics programs as the year 2000 draws closer. Because circumstances and opportunities vary throughout the state, however, a uniform time line would be unrealistic. Regardless of how much students now in school will benefit from new programs, it must be remembered that change takes time and that all elements of change must move in concert. In every school a carefully planned course of action must be adopted, and experimentation and risk taking in the classroom must be allowed. Support for change on everyone's part—teachers, parents, administrators, materials developers, and students—must be nurtured. Only then can positive change take root to the benefit of California's children.

This *Framework* is a vehicle to take us from the present to the future. It does not, however, provide a complete, detailed road map. We cannot be certain what lies ahead, but we must move forward. What the *Framework* asks is that we summon the courage to see where we are and the courage to begin the journey.

Mathematical Power

C h a p t e r 1

In this chapter what is meant by the term *mathematically powerful* is considered, and expectations for the individual student are focused on. The chapter also contains descriptions of what kinds of tasks challenge students to perform at this level and what students need to accomplish those tasks. Finally, the findings of research on how children learn mathematics will be presented to show how students can best be helped to achieve mathematical power.

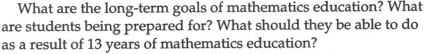

Long-Term Goals

What are the long-term goals of mathematics education? What are students being prepared for? What should they be able to do as a result of 13 years of mathematics education?

All students should be expected to cope successfully with the mathematics they will encounter outside the classroom, including the increasingly sophisticated mathematics demanded in most jobs and most college courses. Also discussed in this *Framework* is the mathematics required for the exercise of competent citizenship in a democracy. Equally important, students should appreciate the beauty and fascination of mathematics and approach the mathematics they will encounter throughout their lives with curiosity, enjoyment, and confidence.

What is the nature of the mathematics students will confront in their lives beyond school? Real mathematics is rarely prestructured or marked with key words. Real situations seldom look like recipes; more often, they are complex and ambiguous. A single task can encompass many problems, often not clearly defined. There may be many ways to go about finding a solution or even deciding what constitutes a solution. Completing a task may take hours, weeks, or even years of sustained, persistent work.

Furthermore, people use mathematics in the everyday world to accomplish a purpose. They seldom work solely to improve their ability to multiply or factor polynomials but rather to accomplish something. Their purpose is not always practical or utilitarian; it may be motivated by curiosity or whimsy or an aesthetic sense.

Finally, although thoughtful work sometimes involves thinking apart from others, real mathematics is seldom practiced in isolation. If the task is at all difficult or requires serious thought, it is best done in collaboration—sharing information, opinions, and expertise. As the task is completed, the results need to be communicated in a clear and convincing way. A combination of individual and group work often characterizes mathematical activity.

It is not enough for students to produce answers to preorganized exercises; they must be able as well to use mathematics to help make sense of real situations. The character of the mathematics that students will confront in their lives beyond school must be kept in mind. Some assignments that students might encounter in a mathematics class are presented here under the title "Exercises, Problems, and Investigations."

Exercises, Problems, and Investigations

A student is asked to calculate:

$3.5 \times 5.5 = ?$

or

What is the surface area of a right square prism whose base is 3.5 cm on a side and whose height is 5.5 cm?

Here the student merely has to follow a procedure correctly to complete the exercise. The task is clear-cut; that is, the student does not have to define the problem, select a technique for solving it, or communicate any thinking. The student has to supply only a numerical answer. The second exercise is more sophisticated than the first only because the student has to solve a verbal puzzle first in order to decide what calculation to do. The expectations for student work are still rather limited. Of course, every student should be able to obtain correct results in exercises like these; but to demonstrate mathematical power, students will have to do much more.

Now the exercise is developed into a *problem*:

Take six rods: one black (length 7); one purple (length 4); and four light green (length 3). Make a three-dimensional object, using all six rods. Find its surface area. Compare your answer with those of other students.

What are the minimum and maximum possible surface areas for objects built with all six rods?

I think the minimum amout is 54 and the maximum amount 102. The one thing I found most is the minimum about of 22. It couldn't be because that is the volume and there is always atleast 3 sides to each rod showing. And 288 is to much because even if all the surface area of all the rods wee showing il would never equal that amount.

Part of a sample eighth-grade response to the problem

Problems like this one are an indispensable part of mathematics education. Here the students must do much more than demonstrate an ability to multiply and add; instead, they must think, visualize, and explain their reasoning clearly. Problems focus student thinking and communication, and the authors have preorganized some of the information and formulated the mathematics with a solution in mind.

Consider the following related task that calls on students to formulate the problem themselves and plan how they will communicate their work:[1]

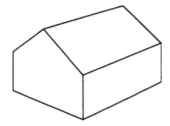

Use paper and tape to build a house having about the same shape as the house drawn above but not necessarily the same size.

Build two more houses, each *similar* to your first house. That is, make them exactly the same in shape but different in size.

Prepare a report describing how you made your houses similar. What can you say, in general, about the relationships among the dimensions, areas, and volumes of similar houses? Glue the three houses to a piece of paper and turn them in with your written report.

Your report will be graded according to your answers to these questions:

1. How well did you explore the relationships among the three houses? How well did you formulate your generalizations?

2. How can it be determined that your houses are similar? Was your approach to making the houses similar a sound one?

3. How well did you present your thinking? Are your ideas understandable to the reader? Did you use mathematical representations, such as graphs, formulæ, diagrams, and tables, effectively?

4. Are your measurements and calculations appropriate and correct?

This is an *investigation*. It is demanding in a way different from a problem and is far more demanding than an exercise. The investigation challenges students to go to the heart of the

[1]Adapted from *Beyond the Surface*. Report of the Investigations Project, designed for and tested on students in grade eight who were not studying algebra. Oakland: Office of the President, University of California, 1991, p. 34.

mathematical idea of proportionality in similar shapes and generalize the results of their mathematical exploration. Investigations, then, can be culminating activities that help students integrate what they are learning; they can also introduce and motivate bodies of mathematics.

Although the exercise requires only a minute or two and the problem perhaps a class period, completing this task might take more than a week in class and outside the class. At the end of the exercise, the student presents an analysis and a conclusion rather than a solution. An investigation is similar to a problem in that students are allowed to use several different approaches. However, the investigation requires that students do more of the formulating of the problem and creates a context that invites more sustained work.

In these examples the problem and investigation both give directions to the students to get them started on the right track. With a problem a student is directed first to build, measure, and compare; but in an investigation an approach is only suggested. Eventually, students will be able to handle a more challenging investigation in which they formulate the issues and problems entirely on their own; for example:

What are the relationships among linear size, surface area, and volume for objects? Make some conjectures and support them with sound mathematical reasoning and concrete examples.

The larger, more encompassing tasks subsume routine work and allow students to demonstrate the full range of mathematical work instead of focusing on its components. For example, to carry out an investigation, students have to understand area and perform calculations; and for this purpose the teacher can show students various computational techniques. Students still learn computation but in a larger, more purposeful context.

In the example just presented, the context of the investigation is mathematical. However, investigations will often be more realistic, asking questions like the following: What is a fair procedure for rationing water in a drought? How are actual measurements on clothes related to the sizes on their labels? How much gasoline do all the cars in our town burn in a year? Investigations can also be more abstract. For example, investigations of strategies for a game, although not everyday mathematics for good citizenship, also require high-quality thinking, depth of mathematical understanding, effective use of tools, and clear communication.

Dimensions of Mathematical Power

When students work on investigations and good problems, they demonstrate their mathematical power. What makes student work mathematically powerful?

Mathematically powerful students think and communicate, drawing on mathematical ideas and using mathematical tools and techniques.

When student work demonstrates all four of these dimensions of mathematical power—thinking, communication, ideas, and tools and techniques—the work is said to be *complete*. (The consequences of expecting complete student work are examined in detail under "Expectations for Student Work," pages 26–32, and under "Assessing Student Work for Mathematical Power," pages 66–71.) In practice students use all of these dimensions of mathematical power at once. For a better understanding of dimensions, each dimension is described separately.

Mathematical Thinking

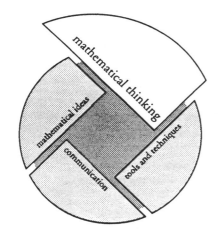

Mathematical thinking at its most powerful grows out of the kinds of thinking that are naturally part of everyone's repertoire. Many of the words used to describe mathematical thinking—such as *classify, plan, analyze, conjecture, design, evaluate, formulate, investigate, model,* and *verify*—have a natural meaning more general than their mathematical meaning. There are other activities we label less frequently in everyday life, such as *deducing, inferring, hypothesizing,* and *synthesizing.* As a group these activities are often referred to as higher-order thinking skills and are characterized in three of the first four standards contained in the NCTM *Standards;* that is: *reasoning, problem solving,* and *making connections.*

Lauren Resnick, a psychologist who has examined in detail how children learn to think mathematically, points out that although defining higher-order thinking skills is difficult, such thinking can be recognized as follows:

Thinking skills resist the precise forms of definition associated with the setting of specified objectives for schooling. Nevertheless, it is relatively easy to list some key features of higher order thinking. . . .
Consider the following:

- Higher-order thinking is *nonalgorithmic*. That is, the path of action is not fully specified in advance.

- Higher-order thinking tends to be *complex*. The total path is not *visible* (mentally speaking) from any single vantage point.

- Higher-order thinking often yields *multiple solutions*, each with costs and benefits, rather than unique solutions.

- Higher-order thinking involves *nuanced judgment* and interpretation.

- Higher-order thinking involves the application of *multiple criteria*, which sometimes conflict with one another.

- Higher-order thinking often involves *uncertainty*. Not everything that bears on the task at hand is known.

- Higher-order thinking involves *self-regulation* of the thinking process. We do not recognize higher order thinking in an individual when someone else "calls the plays" at every step.

- Higher-order thinking involves *imposing meaning*, finding structure in apparent disorder.

- Higher-order thinking is *effortful*. There is considerable mental work involved in the kinds of elaborations and judgments required.[2]

Mathematical Communication

As students do mathematics, they communicate their thinking and understanding to themselves, their peers, their parents, their teachers, and other adults. Students can communicate in many ways: informal conversations, verbal presentations, written text, diagrams, symbols, numbers, graphs, tables, models, and algebraic expressions. Communication helps to clarify a student's thinking, and feedback (formal or informal) can provide useful information for revision. In turn, all students are enabled to improve the quality of their work.

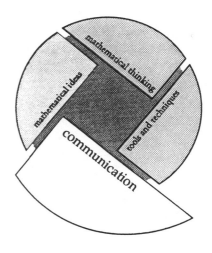

[2]L. B. Resnick, *Education and Learning to Think.* Washington, D.C.: National Academy press, 1987, pp. 2–3.

Besides helping to achieve the purpose for the task, communication provides an opportunity for teachers (and peers) to assess students' thinking and depth of understanding. This communication can be the most concrete product of the student's performance. (This matter will be discussed further in Chapter 2 under "Assessing Student Work for Mathematical Power.")

Mathematical Ideas

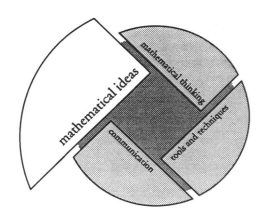

In this *Framework, mathematical ideas* refers to content, the specific subject matter of mathematics as distinct from the other dimensions of mathematical power. Communication, reasoning, tools, and problem solving do not make a person mathematically powerful unless they can be used in conjunction with particular mathematical content. (Content is described in terms of *strands* and *unifying ideas*, which are discussed in greater detail in Chapter 3.)

Strands are familiar subject-matter categories widely used to broaden school mathematics since the 1985 *Framework*. The NCTM *Standards* adopted a similar classification. Using strands helps ensure that programs are sufficiently broad. The use of appropriate ideas from every strand at every grade level helps to maintain a healthy balance.

Unifying ideas are few, ubiquitous, and deep and bridge most or all of the strands. These very important themes of mathematics, such as *patterns* and *proportional relationships*, have a simplicity that makes sense to the youngest thinkers and a depth and power that students can appreciate only after a time. They knit the curriculum together and provide continuity throughout the years. Even so, they are by themselves too abstract and all-encompassing to use as a basis for manageable instructional units. Instead, the mathematics underlying an instructional unit will cut across several strands and unifying ideas. Units will present complex but manageable chunks of mathematics. The collection of units that make up a year's work will develop all of the unifying ideas at that grade level and contain substantial mathematics from all of the strands. (Units are described in greater detail in Chapter 3.)

Mathematical Tools and Techniques

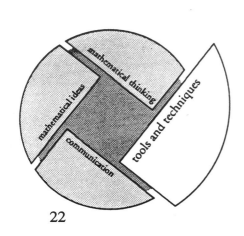

To do complete work, students must put thinking and ideas to use. Involved here are the classic intellectual tools and techniques of mathematics as well as problem-solving tools, including technology. Mathematical tools and techniques enable students to use their hands and eyes to experiment and explore relationships.

In effect these tools and techniques extend the students' thinking power and translate ideas into action.

Students can use tools and techniques to:

1. See patterns and relationships (by using pictures, tables, graphs, blocks, algebraic formulas, or spreadsheet software).
2. Create models (by using drawings, compasses, computer graphic and modeling software, basketballs, toothpicks, and other concrete materials).
3. Determine how many and how much (by using counters, algorithms, calculators, and computation as well as measurement tools such as rulers, scales, paper clips, and stopwatches).
4. Extend memory so that many things can be worked on at once (by using graphs, tables, paper and pencil, manipulatives, and calculators).

The computer is a flexible tool and is ubiquitous in the workplace. For example, students can use word processors to express themselves coherently and legibly, and they can use spreadsheets, graphing programs, and data bases to display results and see patterns. Students can also use Microworlds and modeling programs to ask and answer *what if?* questions.

The ability to select appropriate tools and techniques and use them effectively is an essential part of mathematical power. (The use of technology in the classroom will be discussed in greater detail in Chapter 2, pages 56–60.)

Goals That Support Mathematical Power

Facility with the four dimensions described in the previous section embodies the goal of mathematical power. Three additional goals for students, each important in its own right, provide the underpinnings of mathematical power. The goals are ability to work collaboratively and independently, a positive disposition toward mathematics, and an appreciation of mathematics in relation to history and society.

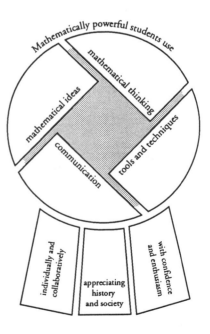

Collaborative and Independent Work

Traditionally, only individual student work and expectations for individual performance have been emphasized in the schools. Looking at someone else's work has been defined as cheating; consequently, many people see mathematics as something done in isolation. They believe that if you want to work with people, you don't do mathematics.

Collaboration is necessary for full participation in society and is expected of workers in the workplace and of citizens in a democracy. Genuine mathematical work, whether done by mathematicians, plumbers, engineers, or dental hygienists, typically involves collaboration together with individual work.

Students receive many benefits from collaborative learning in mathematics. For example, understanding different points of view lets students choose among many strategies. In addition, many tasks are often more interesting and accessible when done with a group. This last point has two consequences. Students in groups are more likely to persist and succeed, and they will be able to handle more sophisticated problems. Both consequences are crucial if expectations for all students are to be raised.

Student collaboration also benefits teachers. Energy that once would have been called disruptive can be channelled into productive communication and problem solving. As teachers observe mathematical thinking in action, it is easier for them to interact with eight groups than with 30 individuals. Collaborative learning provides a structure in which students are given more responsibility for their own learning without teachers losing their responsibility as guides, mentors, and advisers.

In a classroom small-group work and whole-group work complement each other. Small groups are safe; students can take risks and make mistakes. Small groups are also places to learn to listen, discuss ideas, explain, communicate mathematically, and ask questions. On the other hand large groups are ideal for receiving instructions, brainstorming, collecting and summarizing data, seeing different strategies, coming to a common understanding, and summarizing and highlighting important mathematical ideas that arise in the work.

However, individual work and individual accountability are as important as collaboration. Independent work and collaboration flow in both directions. That is, students do independent work to prepare for group work and work in groups to prepare for independent work. Students need to know when it is appropriate to work alone and when to seek out collaborators.

Positive Disposition Toward Mathematics

Students who have a positive disposition toward mathematics are inclined to use mathematics to make sense of situations that come up in their lives and to achieve their own purposes. These students have a good personal relationship with mathematics, which is a part of the way they think about their own lives and the world.

Because mathematical tasks are often complex and ambiguous, students may need to experiment with many different approaches to a task, learn from their mistakes, and try over and over again. To do so requires persistence and hard work. Students who approach mathematics with boredom or fear are unlikely to persist in completing complex tasks. For this reason mathematical power is enhanced by a positive disposition—an attitude of adventure, curiosity, and confidence.

What is the goal for students' disposition? How do ideally disposed students feel and act? They persist in the face of frustrations, erroneous approaches, blind alleys, and difficulties. They enjoy the challenge of a difficult problem and enjoy hearing other people's ideas about how to approach it. They do not want to be told the answer to the problem and can feel a sense of accomplishment discovering something *about* the answer. They have an appetite for finding and explaining patterns and enjoy the power of a generalization. Finally, they are critical of their teachers and the program if they are given assignments without purpose, are left alone, or are given too many answers.

Connection to History and Society

Mathematics has a rich past and present. New materials can expose students to both, connecting the classroom to careers and to various cultures. The goal is that *all* students connect their own lives to the historical and cultural settings in which mathematics has developed and continues to grow.

Students must receive a balanced picture of the range of careers that depend on mathematics and the diversity of the people who have played important roles in mathematics. Historical enrichment can go beyond biography boxes and include cultural connections to mathematics, from the Mayan calendar to Mancala and from the Königsberg bridges to the Towers of Hanoi. Similarly, teachers can celebrate the contributions of diverse mathematicians from all times and places and shed light on the ways in which people from all walks of life use mathematics every day.

Mathematics instruction can be connected to society by a conscious choice of contexts for problems that illuminate the mathematical side of social issues; for example:

The 20 percent of California families with the lowest annual earnings pay an average of 14.1 percent in state and local taxes, and the middle 20 percent pay only 8.8 percent. What does that difference mean? Do you think it is fair? What additional questions do you have? [3]

Such problems take percents, one of the most prosaic work-horses in mathematics, and open them up, breathing new life into them by introducing questions about reporting, statistics, and social justice.

Expectations for Student Work

Simply put, students must be expected to demonstrate mathematical power in their work. That kind of work, *complete* work that demonstrates all of the dimensions of mathematical power, is more elaborate than traditional sets of word problems or other exercises. Mathematical power is too big to fit into a box. Students will do large pieces of work in a wide variety of ways and present the pieces in a wide variety of formats. To make this idea work, the program must:

1. *Foster a common expectation of quality work.* Children and adults must come to expect that large pieces of complete work are part of the program and are to be done well. For example, teachers of history–social science and English–language arts have expected projects for decades. Students also need to understand clearly what is expected and can help set those expectations and standards for assignments in general and investigations in particular.

2. *Prepare students for complete work.* Students need practice in the mechanics of complete work. Practice in writing about mathematics will make their writing clearer and more

---

[3]See "State of Tax Burdens," *USA Today*, April 23, 1991. The figures are averages for a family of four.

concise and practice in presenting will make their presentations better organized.

3. *Give students the time and tools they need to do that work.* Students need a realistic chance to show what they have learned. Large pieces of work take time—to try and err, stall, talk to peers and restart, consider critiques and revise, and mull things over. Much of this time will come outside the class, with class time being devoted to what should be done together. This work also takes tools, both literal and figurative. Students will need calculators, rubber bands, and compasses as well as hints on model building and help with computation.

4. *Require that the mathematical work be done according to a comprehensible standard.* For mathematical power to be exhibited, the mathematics in student work must be clear. Evaluating mathematical thinking is more interpretive than evaluating computational skills. Students and teachers need a common understanding of quality standards, which will be discussed later in greater detail.

5. *Help students develop a habit of draft, feedback, and revision.* All students can regularly prepare drafts of major assignments, get feedback (often from peers), and revise. There is no stigma attached to revision, and there is no grade penalty. Some work will simply need more revision before it meets quality standards. Written work returned to a student for revision should have appropriate feedback from adults or other students indicating what the student might do in order to meet the standards.

6. *Provide tasks worthy of quality large-scale work.* Students cannot be expected to persevere and strive for excellence if the tasks they are given are boring; neither can they develop mathematical power if the tasks are not sufficiently mathematical. Fostering a sense of purpose and mathematical direction in students underlies all curriculum development carried out in this *Framework*.

Setting Standards for Evaluating Student Work

Computational work is easy to score; the answers are right or wrong. But how can mathematical thinking or any of the other large and important components of mathematical power be evaluated? This question falls within the realm of assessment. See "Assessing Student Work for Mathematical Power" beginning on page 66 for information on assessment, including a list of resources.

The assignment, sometimes called the *prompt*, tells the student what must be accomplished. One way to set mathematical standards for large or open-ended pieces of work is to use a *rubric*, the standard against which the work is judged. A rubric is a guide not only to someone evaluating the work but also to the student who wants to perform well.

For evaluation by rubric to be valid and fair, the student and evaluator must share a common understanding of what is to be accomplished in the work. The purpose of the work must be clear in the prompt, and the rubric must parallel the prompt. In general, rubrics should be based on a holistic judgment (taking the work as a whole) of the effectiveness of the work in achieving its prompted purpose. The notion of purpose as used here is tied to the definition of mathematical power in this *Framework*. Specifically, parallelism of purpose as to prompt and rubric should exist in the four dimensions of mathematical power: (1) thinking; (2) mathematical ideas; (3) tools and techniques; and (4) communication.

1. *Thinking.* An *analysis* has a different purpose than a *conjecture* or an *explanation*. In seeking an analysis, the prompt would say, "Analyze the situation." Similarly, the rubric would say, "The analysis of the situation was effective. . . ." Most investigations and problems have a primary thinking goal. In addition, students will and should be expected to use any effective method of thinking to achieve the goal.

2. *Mathematical ideas.* If the student's understanding of geometric similarity is being looked for, the prompt should say so. Thus, students are prompted to describe how the houses were made similar. It is made clear to students that the purpose is not just to make the houses but to make them similar and explain how similarity was accomplished.

3. *Tools and techniques.* The chief interest in the use of tools should be effective usage. Did the tools used serve the purpose well? Were there deficiencies in accomplishment because of the ineffective use or nonuse of tools and techniques? Effective usage clearly entails accuracy and correctness.

4. *Communication.* The primary communication issue in the rubric is how effectively the purpose was accomplished.

In writing, purpose is partly defined by the audience. The prompt should be clear about the purpose and, where appropriate, the audience. Sometimes communication is given too much weight when written work is responded to. Communication is just one dimension of mathematical power, and evaluation should be focused on the reader's understanding of the thinking rather than on other qualities of good writing.

A sample rubric for the "Similar Houses" investigation appears below. It is adapted from a rubric developed cooperatively by ateacher and her students. Here the rubric defines four levels of performance. Many rubrics have six levels, including those used

Scoring Rubric for the "Similar Houses" Investigation

Well done. Your report describes clearly your ideas about the similarity of the houses and how you made them similar. It explains what you did. You explain how you found the size, surface area, and volume of each of the houses in such an elegant way as to make it clear to the reader that you clearly understand the mathematical reasoning behind it. Equations are accurate, clear, and legible. You compare your measurements for different houses clearly and concisely, and you explain the reasoning behind your conclusions about the comparison. The report is organized logically and clearly communicates—through words, charts, graphs, the models, and/or whatever you need—your understanding of the relationships between surface area and volume and your understanding of what makes two houses similar.

Acceptable. Your report does what it needs to do. It includes an adequate description of your method, learning, and conclusions. This report may be less orderly than a *well-done* report, or the relationship between surface area and volume may be less well explained; but you have done the correct mathematics with sufficient accuracy. Yet it is not as clear in its thinking, the ideas you explored, or the use of charts, graphs, or models as a *well-done* report.

Revision needed. Your report may not be complete; or it may lack organization or a clear description of what you did or learned. Your report may show misunderstandings about volume or surface area, about comparing the two, or about similarity; or it may have serious computational mistakes. Your report is being returned to you with suggestions for an *acceptable* or *well-done* report and a time-extension date due.

Restart. Your report misses the mark. You should consider starting over. Perhaps the problem is only in the *report*; but it is so unclear and illegible, has so little mathematics, or has so much missing that it cannot be evaluated. Your report is being returned to you with a time-extension date due. You should see the teacher and your classmates as soon as possible to make sure that you understand fully what the assignment is about and what is expected of you.

---

for open-ended items in tests developed by the California Assessment Program. This rubric allows for flexibility; it does not prescribe exactly what needs to be present but rather lays out what needs to be accomplished. The student must produce quality results in his or her own way. If students help create the rubric, they feel a sense of independence and responsibility. Understanding the rubric compels students to focus on the purpose of the assignment.

It still may take some imagination to connect the rubric to the work. How clearly a student describes his or her ideas about similarity will be evident in the reader's understanding rather than in the student's conformity to a particular scheme. Part of the teacher's job is to understand the evolving relationship between the student and the standards and to distinguish genuine understanding from wishful thinking.

A Sample Response

This section contains part of a student response to the "Similar Houses" investigation. The investigation was the culminat-

---

Introduction
How!

What I did was I first constructed the first House. I then measured the sides. I then found out what the surface area & volume was of the small house. Then I doubled all of the lengths and figured out the surface area volume. The surface area was more than the 1st houses surface area. the volume much larger

also. Then I made my third house. Figured out the surface area and volume Then I compared them all and looked for patterns. That's when I found that you multiplied the first houses block by your number then do it again you get the surface area of your number.

To find the volume you do the same thing exept you multiply it by your number 3 times and you start with the first one's volume.

ing effort in a four-week unit. Written by an eighth-grade girl, the report is nine pages long. A sampling of the girl's narrative and a table she supplied are presented below. The rows of the table beyond 3 are not measurements but predictions based on her discovery.

This girl demonstrates mathematical power. She thinks: predicts, analyzes, and builds a mathematical model; uses mathematical content: area, volume, similarity, proportionality, the mathematics of containers, and scaling; uses tools and techniques: multiplies, adds, uses a calculator, makes concrete representations and tables; and communicates her thinking. Ultimately, it is hoped, her communication will be even clearer, her thinking better organized, and some concepts (such as the number of significant digits) applied more appropriately. In comparison with typical eighth-grade work, her work is exemplary.

Expecting a dozen pages of analysis and supporting data may seem unrealistic for today's students, but expectations must be raised. The girl who did the work just described, although bright, is not exceptional. She was enrolled in a regu-

---

Some mistakes I made was that I tried to build the first house by the pictures you gave us. These measurements weren't correct so I had to start over.

Another mistake I made was on the surface area I would find the one sides surface area then forget to multiply by 2.

T GRAPH

multiplied by how many	length of sides	surface area	volume
1	4.3 cm	73.18 cm²	55.64 cm³
2	8 6 cm	292 72 cm²	445.14 cm
3	12.9 cm	658.62 cm²	1507.87 cm
4	17 2 cm	1170 88 cm²	3560 96 cm
5	21.5 cm	1829.50 cm²	6955 0 cm
6	25 8 cm	2634 48 cm²	12018.24 cm
7	30.1 cm	3585.82 cm²	19084 52 cm
8	34 4 cm	4683.52 cm²	28487 68 cm
9	38 7 cm	5927 58 cm²	40561 56 cm
10	43 0 cm	7318 0 cm²	55640 0 cm
12	5 1.6	10532 92 cm²	96145 92 cm
20	86 0 cm	29272.0 cm²	445120.00 cm

lar eighth-grade mathematics class. The first time she did one of these investigations, she produced a much less sophisticated report. Practice helps. All students can do work like hers, but not overnight. It may take years of similar assignments for teachers and students to become comfortable with expecting mathematically powerful work from all students and confident that high-quality standards are being set and met.

These assignments come not only from the teachers themselves but also from completely new curriculum materials. What should these new materials be like? What do we know about the best ways to present them? Some answers can be found in recent research on learning.

Learning Mathematics

Human beings excel at learning. From birth, children are busy making sense of their world; learning is the result. Children do not have to be motivated to learn; they do it naturally, and they do it well. By grappling with the curious and the confusing, children learn new ways of understanding their world and develop new schemes for thinking.

Children also excel at higher-order thinking. They sing songs, tell stories, and read the tiniest gesture accurately in a wide variety of contexts. It is a myth that children must master lower-order skills before they master higher ones.[4] To do so is not only ineffective but also unnecessary. How else can children sing the songs without decoding musical notation? The problem is not that students are incapable of engaging in higher-level thinking but that this thinking is seldom engaged in the process of learning mathematics.

Over the past decade cognitive scientists and psychologists have looked closely at how children learn mathematics—how they develop higher-order thinking and an understanding of mathematical ideas. This research affirms the concepts Piaget, Vygotsky, and others put forward decades ago. Rather than

---

[4]L. B. Resnick, *Education and Learning to Think.* Washington, D.C.: National Academy Press, 1987, p. 8.

"This assumption—that there is a sequence from lower-level activities that do not require much independent thinking or judgment to higher-level ones that do—colors much educational practice. Implicitly at least, it justifies long years of drill on the "basics" before thinking and problem solving are demanded. . . . Indeed, research suggests that failure to cultivate [higher-order skills] may be the source of major learning difficulties even in elementary school."

—Resnick, *Education and Learning to Think*, p. 8

"Education fails when children are treated as 'blank slates' or 'empty jugs,' ignoring the fact that they have a great deal of mathematical knowledge, some of which surpasses— and some of which may contradict—what they are being taught in school. . . ."

—*Reshaping School Mathematics*, p. 29

being passive absorbers of knowledge, children actively create their own understanding of the world. In fact, by the time they come to school, they have already developed a rich body of knowledge about the world around them, including well-developed, informal systems of mathematics.[5]

Students construct their understanding of mathematics by learning to use mathematics to *make sense of* their own experience. This understanding of mathematics becomes more powerful when students use it to *achieve purposes* that are meaningful to them. Cognitive science and developmental psychology suggest that making sense and achieving purpose are among the strongest organizers of human thinking.

Researchers and practitioners have also focused on the learning of language and writing and the importance of communication in all subject areas. In mathematics, oral and written communication is not only a dimension of mathematical power in its own right; it facilitates the other dimensions as well: mathematical thinking, mathematical ideas, and tools and techniques.

Sense and Purpose

Students come to school with a wealth of experiences; they are experts about their own lives and the culture in which they are growing up. Affirming the ways things make sense outside school and connecting them to things inside is important for students. If students fail to make connections between their personal lives and skills and classroom activities, their access to the curriculum will be diminished.

Connections between out-of-school and in-school experiences increase the likelihood that students of all cultures will connect their out-of-school and in-school thinking. Typically, students get too few examples of the connections between mathematics and the world outside the classroom. A broader range of examples in the classroom provides points of connection; simplified, abstract activities preempt the opportunity for personal connections and thereby diminish access.

Students' experiences in the real world can help the students learn mathematics and see the connections. For example, students can develop the mathematical ideas of *function* and

"All students engage in a great deal of invention as they learn mathematics; they impose their own interpretation on what is presented to create a theory that makes sense to them. Students do not learn simply a subset of what they have been shown. Instead, they use new information to modify their prior beliefs. As a consequence, each student's knowledge of mathematics is uniquely personal."

—*Everybody Counts*, 1989, p. 59

---

[5]*Reshaping School Mathematics: A Philosophy and Framework for Curriculum*. Washington D.C.: National Research Council, 1990, p. 29.

variable as extensions of ideas they already have about growth, motion, or cause and effect. Even young children search for patterns in their everyday lives, constructing concrete functions. The challenge to educators is to capture this natural inclination and transform it into mathematical understanding. For example, students might be led through the following:

You are planning to give two cookies to each friend who comes over after school. Make a drawing to show how many cookies are needed for one friend; for two friends; for three, four, and five. Can you see a pattern? How can you say the pattern? Will the pattern keep going? Show the pattern with numbers.

Cookies are a good choice for the problem because they are familiar to students. That statement may be true, but the point to be understood is not that sharing cookies helps students understand mathematics but instead that *mathematics helps students understand how to share cookies.* When mathematical ideas help students understand situations, the ideas are not only more interesting and accessible but also easier to remember and use.

As students get older, they can handle more complex problems. For example, by the middle grades students have had plenty of experiences waiting in line—at the movies, in the cafeteria, and in stores. With these experiences as a starting point, students can be led to speculate on general relationships among waiting time, the number of people in the line, the position of the end of the line, and the length of time each person takes to buy a ticket or a lunch.

The exploration of these relationships, empirically and theoretically, helps students not only to make sense of the situation but also to achieve a purpose. For example, students might answer this question: What is the most practical way get everyone through the lunch line more quickly? Students can use data collected from the line to make mathematical models that can help them predict waiting time and evaluate their suggestions.

Materials and situations are used that give context and purpose to the assigned activity; in turn, the context and purpose give meaning and organization to students' thinking. A framework for learning is provided by capitalizing on students' natural disposition to think and reason toward achieving a purpose.

As students mature, they can reason more abstractly. They can quantify and symbolize more complicated relationships in the situations they investigate. For example, in the Tower of Brahma (also known as the Towers of Hanoi) puzzle, how many moves does it take to move a stack of *n* disks from one peg to another? A solution to the puzzle can lead to a proof that it takes $(2^n - 1)$ moves. Students are not expected to connect the puzzle to their everyday lives in order to want to solve the problem. Instead, their sense of purpose is expected to come from an interest in mathematics for its own sake and, in turn, comes from past experiences consciously connected to students' everyday life: their needs, language, and culture.

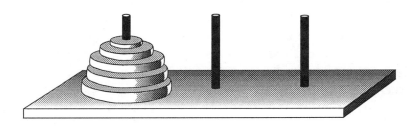

The Tower of Brahma puzzle: How many moves does it take to move n disks from one peg to another?

- There are *n* disks on pegs.
- Move only one disk at a time.
- Never put a larger disk on a smaller one.

Where does this puzzle come from? According to legend, in the temple in Benares, India—under a dome that marks the center of the world—there is a setup like this with 64 disks that priests move day and night. God placed the 64 disks on one needle; when they have all been moved to the next (following the rules, of course), the universe will end.

If they are right, how long do we have?

Mathematics, Language, and Communication

Mathematics is a special language that is learned by use. All students benefit from a regular diet of mathematical talk, writing, drawing, graphing, symbols, numbers, and tables to help them think and communicate their ideas. By making sense to others, they make sense for themselves.

Mathematical language and everyday language are interrelated, as are mathematical thinking and everyday thinking. They overlap and reinforce one another. If students practice thinking and communicating mathematically and connect their day-to-day experiences, they will develop a greater repertoire of techniques in both arenas. They might be more systematic in exploring possibilities outside the mathematics class and more

"[Students construct their own mathematical understanding] most readily when students work in groups, engage in discussion, make presentations, and in other ways take charge of their own learning."

—*Everybody Counts*, pp. 58–59

likely to use common sense inside. Working with mathematics to communicate helps all students—whether they are fluent in English or another language or are developing proficiency—benefit from dealing with the precision of speech and writing that good mathematics demands. Experience from language arts and research in cooperative learning and writing in mathematics introduce more issues. Many writing specialists believe that a command of diction, sentence structure, and grammar develops better when students write for a purpose and an audience rather than work on isolated exercises. The same approach is valid in mathematics as well. Children cannot learn to communicate in isolation because when they communicate, they communicate with others. That communication is most effective when students express themselves mathematically for a clear purpose and audience.

A working example of meaning, purpose, and audience

The teacher assigns each student to choose something wanted (e.g., a bicycle, a tape deck, a trip to Latin America) but cannot afford. Work begins with research—trips to stores, to agencies, catalog shopping, and so forth—to collect data: How much do these things cost? Is there much variation from store to store? From vendor to vendor? Are there options or variations in quality that affect the price? After the research is completed, students give a written report or oral presentation to their group. The teacher helps students explore the mathematical elements of their choices and decision.

Next, students develop a plan for saving the money to get what they have reported on. They brainstorm savings schemes, from piggy banks to interest-bearing accounts, and money-making ideas, from lemonade sales to summer jobs. The students now have a reason to assimilate the mathematical concepts they need to evaluate the relationships between cost, savings, and time required. The concepts become part of the students' natural thinking; they construct durable knowledge that corresponds with what they already know. They use the mathematics, intertwined with common sense and experience, to complete a report. The report is a proposal to someone to lend them the money; their savings scheme serves as the plan for how they will pay the money back.

The students produce a proposal for a committee of students or an outside guest, whom they attempt to persuade through the strength of mathematical reasoning. In preparation the teacher leads students to use a range of mathematical representations—words, tables, diagrams, and formulae—in a specific context. Other students respond to drafts of the report and to the presentation. Students revise and edit their reports. By the time the final report is produced, everyone has focused on issues of what defines quality and insight in mathematical work.

Andrea   8th

I think that writing down your thoughts is a good way to get them out in the open and helps me sort through the material I've learned. I have found that writing things down helps me to think more clearly. That's why I like to do this so I understand better.

Comments of an eighth-grade mathematics student on value of writing

The audience can be real or imaginary: teacher, peer, parent, outsider, or self. Communication can be formal or informal, written or oral. Students will learn how to convey more information and construct better arguments by being assigned to spend time in class and out talking with each other, responding to the written work of others, and investigating new situations and ideas collaboratively. As their facility for expressing their own understanding develops, their ability to understand the ideas of others grows as well.[6]

Learning and New Programs

What has been gained from the previous discussion that can be used to develop new mathematics programs? Here are several principles:

- Making sense and achieving purpose are fundamentally important in students' learning of mathematics.
- Every student has access to higher-order thinking. The key is to unlock the world of mathematics through a student's natural inclination to strive for purpose and meaning.
- Communication—with themselves, with peers, with parents, with other adults, and with the teacher, orally and in writing—helps students learn mathematics as they clarify their own ideas and hear ideas from others.

---

[6]Methods of communication-based learning are effective for foreign language learning and models of instruction in English as a second language. See, for example, the *Foreign Language Framework* (Sacramento: California Department of Education, 1989).

- Mathematical language itself is a thinking tool that facilitates mathematical understanding and connects to natural language and everyday thinking.

Beyond these principles there are some other implications of research on learning that deserve special mention:

Quality work takes time. When students are engaged in producing something that makes sense, with ample time to reflect and revise, they are more likely to succeed. Allocating enough time tells students that reflection is more valued than reaction; the quality of the final product is more important than the quality of the first draft; and the ability to make use of all of the available resources—including classmates—is more important than bare-handed mathematics. If the program supports and rewards persistence, more students will conclude that they should try.

Deep understanding takes time. The most important understandings in mathematics grow continually over a student's career. While the idea of *mastery* might make sense when talking about shallow skills, the big ideas of mathematics (discussed in Chapter 3) continue to deepen with experience and maturity. Students need to revisit central ideas like *functions* over many grades and under different circumstances. Building on informal examples in elementary school (how many cookies for how many friends?) and more explicit modeling in middle grades (how long will we have to wait in the cafeteria line today?), older students are able to raise their personal experience and more sophisticated thinking to more abstract mathematical understanding.

Developing Mathematical Power in the Classroom

C h a p t e r **2**

Chapter 1 was focused on the development of mathematical power from the perspective of the individual student. Chapter 2 is concerned with the development of mathematical power from the perspective of the classroom. The essential characteristics of an empowering mathematics program are described briefly, then explored in depth, and finally examined as to how they might appear in the classroom, traditional practices being compared with those recommended in this *Framework*. (The structure and mathematical content of empowering programs are discussed in Chapter 3.)

Characteristics of Empowering Mathematics Programs

A mathematics program consists of several elements, none of which are discrete or independent of the others:

- Materials students use: textbooks, supplemental materials, manipulatives, calculators, computers, and other tools, such as compasses
- Tasks, problems, and assignments on which students work
- Expectations for how students are to work: the extent of independent and collaborative work, the amount of talking permitted or required, and the speed and depth expected for their work
- Teaching methods: lecturing, assigning seatwork or investigations, and questioning or probing students' thinking
- Assessment: tests, teacher observations, group presentations, or students' writings about their work

In an empowering mathematics program, *students learn mathematics.* The program provides the written and physical materials, the pedagogical environment, and an overarching philosophy that supports the teachers and students. By integrating the characteristics described in this section, the program helps all students learn a comprehensive body of meaningful mathematics. The characteristics are ineffective if kept independent of one another; they must be brought together to produce results. An empowering mathematics program at all grade levels, kindergarten through grade twelve, shares the following essential characteristics:

1. *All students participate fully.* An empowering program assumes that students come from a wide range of cultural, economic, and linguistic backgrounds. Tasks and problems will be accessible to all students and yet will be rich enough and open enough for interested students to investigate beyond standard expectations. With this support all students can be expected to pursue mathematical interests, express their ideas, and learn from one another. In an empowering program no students will have to do "simplified" material; neither will they be held

back from challenging ideas because the content is slotted for a later grade. Keeping all students in the core program is especially important because students in low groups or tracks are disproportionately nonwhite and come from language-minority groups. The program also takes advantage of student diversity to make material accessible by using students' own experiences to help create the specific contexts for working with mathematics. (See "Gift of Diversity," p. 44, and "Grouping Students," p. 60.)

2. *Students take responsibility for their learning; they question, create, and help decide what to do.* The program demystifies mathematics education and destultifies students. It makes sense to students and helps them make sense of their world. Students are given active roles in their own education in familiar and interesting contexts. The power in empowering programs does not come just from mathematics. By being given control and responsibility, students are also supported in their sense of purpose. (See "Tasks with Purpose," p. 46, "Learning Mathematics," p. 32, and "Grouping Students," p. 60.)

3. *Teachers are facilitators of learning rather than imparters of information.* They encourage and support students as they become active learners; establish a classroom climate in which students construct their own understanding of mathematics, providing rich environments for student investigation; and facilitate learning by asking questions that probe students' understanding and provoke their inquiry. In addition, teachers ask students to explain their thinking, help them to hear and react to the ideas of others, and encourage them to make and refine mathematical generalizations. (See "Role of the Teacher," p. 49.)

4. *All students regularly use manipulatives, calculators, and computers.* Students have access to a wide variety of resources, as they would in the workplace. Throughout their school careers they use manipulative materials and technology to explore mathematical ideas, model mathematical situations, analyze data, calculate numerical results, and solve problems. They make choices about what tools are needed and when to use them; and they are in control of the technology, determining branches to take, routines to call up, or programs to write. They use manipulative materials to illuminate concepts, model mathematical

situations, and solve problems. (See "Manipulatives and Technology," p. 56.)

5. *All students frequently work together, sharing and discussing ideas.* Although students regularly work alone, they also regularly interact in small heterogeneous groups. They grapple with the mathematics, expressing their own thoughts, comparing alternative approaches, sharing insights, and testing conjectures. They state their own arguments and listen carefully to others. There is a genuine give-and-take among students and with the teacher. The class establishes and lives up to expectations as to how to share listening time and how to give quality responses to peers. (See "Grouping Students," p. 60.)

6. *All students frequently reflect their thinking orally and in writing.* Students talk and write frequently, reflecting on what they have done, clarifying their own thoughts, and communicating their ideas and results with others. They organize and record their accomplishments and findings and give evidence or arguments supporting their conclusions. They incorporate diagrams, mathematical symbols, and natural language (including non-English) in their explanations. Questions of *why* and *how* permeate the program. Students critically examine their own solutions and interpretations as well as those of others. (See "Grouping Students," p. 60; "Tasks with Purpose," p. 46; and "Assessing Student Work for Mathematical Power," p. 66.)

7. *Assessment is integrated with instruction; it focuses on what students understand and can do rather than on what they don't know or can't do.* Designed to reveal students' mathematical power, assessment points students towards success rather than failure. Most assessment tasks have the same depth as learning tasks, including a balance of exercises, problems, and investigations. Students do mathematics: formulate problems, consider and apply a variety of approaches, and determine and explain their findings. They use tools, such as calculators and manipulatives, as freely in assessment as they do in instructional tasks; and they have ample time for their work, including time outside the class, and access to their notes when being assessed. Finally, they have the opportunity to revise and

resubmit important assignments to bring performance and grades up to high-quality standards. (See "Assessing Student Work for Mathematical Power," p. 66.)

8. *The program is appropriate to the maturity and development of the students as it meets its other goals.* The program recognizes that students of different ages learn differently and have different problems; therefore, every part of the curriculum is appropriate to its grade level. The program takes into account the psychological difficulties inherent at any age level and considers the major mathematical ideas at a level that the students' development makes understandable. As students mature, they investigate these mathematical ideas more and more deeply, visiting the central and most useful ones again and again in different contexts. (See "Special Considerations for the Primary Grades," p. 72.)

9. *The program develops every student's positive disposition toward mathematics in several ways.* How can all students be helped to approach mathematical tasks with persistence, confidence, and enthusiasm? First, the broad scope of investigative work, the intellectual challenge of problem solving, and the effort required to do both reward persistence. Second, creating a cooperative setting and valuing the variety of possible approaches encourage confidence and enthusiasm in students. They will come to believe that effort, sensibly applied, produces the best work and that "being good at math" is not simply a matter of ability. (See "Developing a Positive Disposition Toward Mathematics," p. 71.)

10. *The program usually introduces computational procedures only when students need them.* Students use computational procedures as tools, not as ends in themselves. They usually develop facility with particular techniques, ranging from two-digit subtraction to integration by parts, in the service of solving problems in appropriate contexts. Teachers encourage students to invent and explain their own procedures and to compare different procedures to see which is best in a particular circumstance. Finally, understanding is always preferred to speed. (See "Procedures for Manipulating Numbers and Symbols," p. 54, and "Acquisition of Number Facts," p. 74.)

The Gift of Diversity

California's school population is changing, and along with the change in the population itself is a change in attitudes. Instead of diversity being viewed as a challenge, it can now be seen as a gift. California has discovered a new vein of gold, a rich environment that includes many points of view, many languages and cultures, and many approaches to sharing and learning in our classrooms.

Along with the gift of diversity come responsibilities. Empowering mathematics programs are inclusive; they use nonracist and nonsexist language, culturally diverse situations, and teaching materials that make mathematics accessible. Such programs attend at every instant to the idea that mathematics is for everyone in the classroom and in the school—that all students should feel that, no matter who they are, their mathematical goals are attainable.

Gender, Race, Class, and Culture

Regardless of the causes of differences in interest and performance in mathematics related to gender, race, class, or culture, efforts must be continued to weed out the stereotypes common in family, school, and society. Some current practices, typically unintended and unconscious, play a role in increasing differences. For example, teachers unconsciously tend to treat boys and girls differently in mathematics classes; that is, teachers give boys a disproportionate amount of time and attention. (*Note:* The increased attention that teachers give to Latino, African-American, and low-income males in mathematics classes is more likely based on disciplinary than academic reasons.)

And students accept the situation. They, too, play out roles in the insidious drama of life at school. Change is difficult because the participants are unaware of their roles in these transactions. Sensitivity, patience, and moral stamina will be required of everyone for mathematics to become an instrument of equity. This failure is not new; some students have always fallen through the cracks. The changes that have to be made now to take full advantage of diversity will enrich us all.

"Asian teaching practices thrive in the face of diversity, and some practices even depend on diversity for their effectiveness. . . . While American schools attempt to solve the problems of diversity by segregating children into different groups or different classrooms and by spending large amounts of regular class time working with individual students, Asian teachers believe that the only way they can cope with the problem is by devising teaching techniques that accommodate the different interests and backgrounds of the children in their classrooms."

—Stigler and Stevenson, "How Asian Teachers Polish Each Lesson to Perfection," 20

Language-Minority Students

Language is necessary to the learning of mathematics. It bridges new understandings and a student's previous knowledge and seals them. Students learn mathematics best in their primary language; therefore, they must be given the opportunity to do mathematics and create their own meaning by speaking, writing, and reading mathematics in their primary language.

Students developing proficiency in English can make significant contributions to mathematical understanding for all students. Some advantages are obvious: counting in other languages, introducing multicultural situations, and teaching games from other cultures. Others may require more awareness. For instance, expressions such as *one more than* or *half as many* may have different meanings or no meaning at all in other cultures or languages. The adventure of defining the expressions and discussing them can broaden the understanding of those concepts for both students and teachers.

Ideally, all students would be taught by qualified mathematics teachers who speak the students' native languages. Unfortunately, these teachers are in short supply. Alternative strategies include teaming with a teacher who speaks the language, asking bilingual students to help, and learning more about how to help students acquire a second language. Lessons can also be made more accessible by using many visuals, showing and demonstrating, and building vocabulary. At the same time mathematical ideas must not be simplified for students whose native language is not English. These students need access to all levels of mathematics, not just a standard mathematics class for language-minority students.

Steps That Can Be Taken

What can be done to take advantage of what diversity offers? Some suggestions in several categories are as follows:

1. *Curriculum and assessment.* Use contexts and representations from other cultures. Provide opportunities for presentations in other languages and activities that require materials familiar to special groups. Many students can, for instance, demonstrate mathematical concepts through their skill in using money.

2. *Staff development.* Arrange for opportunities for teachers to expand their understanding of student differences and other cultures through in-service training and sharing.
3. *Quantity and quality of instructional time.* Make sure that not only equal time but equal quality as well are given to all students. Arrange for peer observations or videotaping for self-checking.
4. *Grouping.* Maximize time spent by students in heterogeneous groups. Minimize time spent in tracked or special-ability groups.
5. *Beyond the bells.* Pay heed to the time students spend outside school. Arrange times and places for study groups to do homework. Have students exchange phone numbers so that the students can discuss homework problems. Involve the community (parents, business, local colleges) in organizing supportive settings for students to study, work on projects, or do homework. Provide programs such as "Family Math," in which students and their parents understand and enjoy mathematics together in the same way in which they might enjoy reading together.
6. *Community and parents.* Call on the many resources available from members of the community and parents, such as translation or language interpretation, career information, and assistance provided by mentors.
7. *Systematic support.* Meet with all groups involved, including students, and plan systematic support. Evaluate, revise, and update the plan regularly.

A large part of the mission will be giving up the norm: how students are viewed, what they are to be offered, and what gains are expected of them. The issues are complex, involving the social context of the classroom, the general culture and languages, culture and languages specific to mathematics, influences outside school, and the need for teacher to have opportunities to learn more. The new norm will be achievement in mathematics for all students.

Tasks with Purpose

The signature of empowering mathematics programs is interconnected, worthwhile tasks that include good problems and investigations, such as those described in Chapter 1. Many tasks will mirror the complexity of life situations; others may be purely mathematical investigations. In either case students who work on them do so with a sense of purpose. The work they do will not be work for its own sake but work to accomplish something meaningful to the student.

Teachers and curriculum designers benefit from thinking about the student's sense of purpose. Mathematical tasks have traditionally avoided a purpose connected to the students; the purpose of exercises has always been to develop facility for the mathematics coming next. But purpose gives order to mathematical activity, just as it gives order to writing. A student's purpose need not always be utilitarian ("I want to predict how much wood I need to build this shed"). It can also be whimsical or aesthetic ("I want to know how many Ping-Pong balls will fit into the cafeteria." "I want to see whether I can build an icosahedron from straws and string."). When students want to perform well, they can help set the standards and take part in evaluating each other's work. Making the purpose *theirs* removes a layer of mystery surrounding the correspondence between learning and percent right.

For the development of a sense of purpose for all students, tasks are selected for their richness and complexity, their contribution to the coherence of a unit, and the opportunities they provide for students to do complete mathematical work. Accessibility to a broad range of students is an essential consideration. The tasks are open, are accessible in many ways, and, more often than not, have multiple solutions. They may appear in sequence or in a menu format whereby students decide in what order the tasks will be done. There may be individual tasks, but small-group work should be commonplace. Students may generate questions about a situation, define problems to explore, find a solution or solutions, and prepare a report or presentation of their results.

These tasks require time and deliberation; many will take place over several days. Well-designed investigations have no end points; students delve into questions to a lesser or greater depth depending on their interests. Quick answers, ten-minute problems, and short-term mastery of mathematical ideas are not enough.

The items on page 48—all related to students tasks—are taken from a complete chart titled "Traditional, Alternative, and Desired Classroom Practices," which can be found in Appendix C. The middle column, "Some Alternative Practices," is not an intermediate or necessary step but illustrates what some teachers have been doing since the 1985 *Framework* was issued. Other items from the chart appear throughout this chapter and in Chapter 3.

"The standards specify that instruction should be developed from problem situations. As long as the situations are familiar, [concepts] are created from objects, events, and relationships in which operations and strategies are well understood. In this way, students develop a framework of support that can be drawn upon in the future, when rules may well have been forgotten but the structure of the situation remains embedded in the memory as a foundation for reconstruction. Situations should be sufficiently simple to be manageable but sufficiently complex to provide for diversity in approach. They should be amenable to individual, small-group, or large-group instruction, involve a variety of mathematical domains, and be open and flexible as to the methods to be used."

—*Curriculum and Evaluation Standards for School Mathematics*, p. 11

	Traditional practices	Some alternative practices	Recommended practices
Assignments ▶	Daily assignments are given from the textbook. They are expected to be done individually. "Drill and practice" assignments from the textbooks are given frequently.	Students use manipulatives and work in small groups, but fairly prescriptive work sheets often guide students' work. There may be a project, but the mathematics behind it may not be evident from the product.	Assignments help students develop mathematical power; they are challenging and multidimensional. They may extend over several days and require considerable time outside the class. Often, they are cast in the form of a broad, complex problem to investigate as a group. Findings or results may be presented in an individual or group report, an oral presentation, or visual displays.
Lessons ▶	A two-page lesson per day is organized around a specific objective. The lesson includes: • Application example • Instruction in how to do a specific procedure, perhaps with some simplified explanations • A few exercises to check for understanding • Practice exercises • Optional enrichment problem	In a one-day or two-day lesson, work sheets provide specific directions for the student on what to do and how to record results.	Most lessons are multiday and generally involve more than one big mathematical idea. Students work with important ideas over an extended, continuous period, sometimes as long as six weeks. Mathematical ideas that arise in the lessons are the subject of student reflections and teacher-guided summarizing discussions.
Review ▶	Approximately the first third of the year is spent reviewing computation and symbolic manipulation topics previously taught. New material is presented in chapters but is reviewed only on semester tests.	Review is provided during ten-minute warm-ups at the beginning of the period or in "sponge" activities done sporadically. Occasionally, students work on problems that incorporate work they have studied from several strands.	Review is built into the rich and complex problems students work on.

Role of the Teacher

Teachers who contributed to the development of the *Framework* view the role of the teacher as follows:

> As teachers we would like to establish in the classroom a climate in which students take responsibility for understanding, doing quality work, managing their time, and communicating with others. Our classrooms can be student-oriented, self-directed, and nonauthoritarian. We can drop the role of a figure who passes judgment on what is right and wrong. We frame questions, plan work that excites curiosity, and encourage our students to exploit what they know and intuitively feel about the situation at hand. Within definable instructional objectives, we assume the role of guide and pacer and conduct the students into regions uncharted for them.

The perspective of teachers that is developed in this *Framework* includes their view of the role of practicing mathematician:

> As we assume this new role, it is important that we, too, take on the role of practicing mathematician. Our students need to watch as we attempt to solve problems we do not have answers to. They need to see us struggle with messy situations, persevere, conjecture, and test our theories. In this way students will realize that finding a solution is not always easy but can be exciting. In short, we need to model the mathematical behaviors and positive disposition we want in our students.

The 12 guiding principles of teaching for understanding in the *Mathematics Model Curriculum Guide, Kindergarten Through Grade Eight*, hold true for all teachers of children in kindergarten through grade twelve. These principles, reprinted on pages 52–53 of this *Framework*, are fundamental to empowering mathematics programs.

Facilitating and Supporting Student Learning

The classroom is an intellectual environment in which students and teacher engage in serious mathematical thinking. Teachers continually reflect on questions like these:

1. How can I structure the classroom environment and my interactions with students so that they want to confront and make sense of mathematics?
2. How can I understand what the students understand?

> "The ideal classroom fosters the spirit of 'discovery.' It also provides a variety of ways for pupils to direct their own learning under the mature, patient guidance of an experienced, curiosity-encouraging teacher."
>
> —*Mathematics Framework*, 1972, p. 7

> "Group work changes a teacher's role dramatically. No longer are you a direct supervisor of students, responsible for insuring that they do their work exactly as you direct. No longer is it your responsibility to watch for every mistake and correct it on the spot. Instead, authority is delegated to students and to groups of students. They are in charge of insuring that the job gets done, and that classmates get the help they need. They are empowered to make mistakes, to find out what went wrong, and what might be done about it."
>
> —Cohen, *Designing Groupwork*, p. 91

Because hands-on experiences help students make sense of mathematics, a variety of resources must be provided for students to use. Calculators, computers, and manipulative materials are available when the students want or need them. Teachers need to incorporate tools and technology in their lessons and assignments.

Teachers need to organize their classrooms so that students think and talk about their work. Students share ideas and question other students as well as the teacher. This communication takes place when students work in pairs or small heterogeneous groups on tasks or investigations. Students will be expected to share approaches, conjectures, difficulties, and results within their group and among groups. Informal instruction from peers often handles incidental gaps in understandings or techniques and hones the communication skills of the more knowledgeable students. However, teachers need to exercise caution. To change overnight from rows and lecture to small heterogeneous groups would be precarious and probably counterproductive. Both teachers and students will need preparation and practice to make the change.

Not all the teacher's time should be spent in groups. Class time for core activities in the curriculum is typically spent in a combination of heterogeneous small-group, whole-class, and individual work. Whole-class work includes teacher-guided discussions, sharing across groups, student presentations, lecture with active student participation, and summarizing.

The teachers involved in the development of this *Framework* describe themselves as follows:

> As part of our teaching role, we observe students as they work and interact with one another. We listen to students and question them to understand what they are thinking. Asking the students to explain their work not only helps us to learn about the student but also helps the students clarify what they are doing and what they understand. We probe students' thinking, pushing them to think critically about their work.

When students can make sense of an explanation, when it fits in with what they already know and understand, teachers share their own knowledge with them. Students understand and use a mathematical explanation best when they have personally grappled with the issues, identified questions, struggled, and reached conclusions. Then they are ready to understand how others explain the same situation.

Traditional practices	Some alternative practices	Recommended practices
Most of the time students are expected to work alone. They sit in individual desks, facing the front of the room, although primary students may sit at tables.	Most of the time students are expected to work alone. For problem-solving activities, students arrange their desks in groups of four.	Students sit in small heterogeneous groups. They are encouraged to interact with each other. For many assignments students may choose to work alone or with others.

◀ **Student interaction**

Making Instructional Decisions

As teachers come to know their students by working with them in a variety of activities, instructional decisions are guided by questions such as the following:

1. Is the work challenging as well as accessible to all students? Do all the students understand the task and know how to proceed?
2. Is each student contributing to the work of group tasks?
3. What are common misperceptions? How are the students dealing with them? If a misperception is serious, is intervention necessary? Or will students discover the misperception by themselves? How can a question help them uncover it?

The students keep responsibility for thinking and doing, and teachers do not always rescue them with ready-made remedies for their problems. Sometimes students need to flounder so that the groundwork can be laid for new understandings. In many cases the best instructional decision is not to interfere. At other times teachers may decide to intervene directly with a question or clarification to the student or small group or to intervene indirectly by focusing whole-class discussion on an apparent difficulty or to revise the plan for the next day.

Teachers need to think carefully about decisions regarding feedback to students. When is feedback important? How can feedback best support and encourage students to think critically about mathematics? Sometimes teachers give direct feedback to students, and at other times they question students to help the students recognize what they have accomplished or are con-

Teaching for Understanding: Guiding Principles

1. Our top priority should be the development of students' thinking and understanding. Whenever possible, we should engage the students' thinking and teach the mathematical ideas through posing a problem, setting up a situation, or asking a question.

2. We must know that understanding is achieved through direct, personal experiences. Students need to verify their thinking for themselves rather than to depend on an outside authority to tell them whether they are right or wrong. We must see our job as setting up appropriate situations, asking questions, listening to children, and focusing the attention of the students on important elements rather than trying to teach a concept through explanations.

3. We must know that the understandings we seek to help the students gain are developed, elaborated, deepened, and made more complete over time. We must provide a variety of opportunities to explore and confront any mathematical idea many times.

4. We will not expect all students to get the same thing out of the same experience. What students learn from any particular activity depends in large part on their past experiences and cognitive maturity. We should try to provide activities that have the potential for being understood at many different levels.

5. To maximize the opportunities for meaningful learning, we should encourage students to work together in small groups. Students learn not only from adults but also from each other as well.

6. We must recognize that partially grasped ideas and periods of confusion are a natural part of the process of developing understanding. When a student does not reach the anticipated conclusion, we must resist giving an explanation and try to ask a question or pose a new problem that will give the student the opportunity to contemplate evidence not previously considered.

7. We must be interested in what students are really thinking and understanding. Students may be able to answer correctly but still have fundamental misunderstandings. It is through the probing of the students' thinking that we get the information we need to provide appropriate learning experiences.

8. We must be clear about the particular idea or concept we wish students to consider when we present activities or use concrete models. It is not the activities or the models by themselves that are important. What is important is the stu-

dents' thinking about and reflection on those particular ideas dealt with in the activities or represented by the models.

9. We need to recognize that students' thinking can often be stimulated by questions, whether directed by the teacher or other students. We should foster a questioning attitude in our students.

10. We need to help students develop persistence in solving problems. Only in a learning environment in which mistakes and confusion are considered to be a natural part of the learning process can students believe they do not have to come up with quick, right answers.

11. We need to recognize the importance of verbalization.

Putting thoughts into words requires students to organize their thinking and to confront their incomplete understanding. Listening to others affords them the opportunity to contemplate the thinking of others and to consider the implications for their own understanding.

12. We must value the development of mathematical language. Language should serve to internalize and clarify thinking and to communicate ideas and not be an end in itself. Memorizing definitions without understanding interferes with thinking. The emphasis is on developing a concept first, establishing the need for precise language, and then labeling the concept accurately.

Mathematics Model Curriculum Guide, Kindergarten Through Grade Eight, p. 13

fused about. Teachers realize that more is gained by recognizing accomplishments than dwelling on mistakes, that specific feedback is usually more useful than global praise or dismissal, and that it is more effective for students to discover their own mistakes than to have them pointed out by others.

When teachers respect and celebrate the diversity of the classroom and students follow the teachers' lead, the students also learn to appreciate the variety of ideas, approaches, strategies, and solutions to a problem. They come to respect everyone's work, producing a classroom climate that encourages students to take risks and venture incomplete thoughts, knowing that others will try to understand and will value the individual's thinking.

Procedures for Manipulating Numbers and Symbols

How do computational procedures fit into this mathematics program? Students should be able to find answers to problems that call for computation, and they should know the basic facts. The same can be said for symbolic manipulation. Older students should to be able to solve equations and recast expressions, using their facility with algebraic manipulation.

Children often misinterpret and misapply arithmetic and algebraic procedures taught the traditional way. This program, in contrast, values developing number and symbol sense over mastering specific computational procedures and manipulations. The goal is that students calculate accurately when asked and be able to assess whether their answers are reasonable. Similarly, older students should develop symbolic understanding so that they can translate from situations to equations and back again. Students should know which operations or manipulations are called for and to determine which method of calculating will be most efficient a given situation.

Because students will still perform calculations in many tasks, they will have to determine which numbers, symbols, and operations are appropriate. They will calculate in their mind only or with the aid of a calculator or paper and pencil, choosing which is best for the task. And they will check to see that their answers are reasonable.

Estimation is relevant here because some calculations do not require precise answers. In fact, part of the student's job may be to determine the precision required by a calculation. Therefore, estimation is not solely a means of checking required by calculation; it may be the appropriate technique in itself. Checking for reasonableness is not confined to numbers. Part of symbol sense is being able to tell whether a symbolic result is plausible—is based, for example, on the powers of the variables or on dimensional considerations.

There are many different ways to perform a calculation. Some students will invent and refine procedures themselves, and immigrant students may bring valid alternative procedures. Teachers can encourage inventions and other alternatives and help students share them with others.

On the other hand students should not be expected to invent or discover specific mathematical algorithms or procedures. The latter are good and efficient, but their usefulness becomes clear only after a person has confronted many problems for which the procedure can be used. Teachers should not be reluctant to demonstrate or explain a mathematical procedure to students when they can see why it is useful. Then they will easily incorporate the procedure with their own work. Students

---

"How can number sense be defined? . . . It turns out to be nearly impossible, because number sense, upon reflection, is not a collection of things that one knows about numbers or of skills that one can exercise upon numbers. Rather, it is a set of not fully predictable things that one tends to do with numbers under certain circumstances on the basis of a body of interrelated concepts of number and knowledge of specific numbers.

The following is a list of possible indicators (not components) of number sense:

- Using well-known number facts to figure out facts of which one is not so sure. Note that this will be observed only when the particular individuals involved feel they know the "benchmark" number fact much better than the fact to be figured out, when they value accuracy more highly than speed, and when there is no more efficient procedure (such as counting) available to them.
- Judging whether a particular number constitutes a reasonable answer to a particular problem. Note that someone might be able to judge what is reasonable in some situations, but not in others. Note also that we would normally see this kind of number sense only when a person has generated a wrong answer.
- Approximating numerical answers (rather than calculating exact answers). Note that this will be used only when the individual judges that an approximation is adequate and when it is easier (quicker, more reliable) for that person to approximate than to calculate exactly. The latter may depend, in part, on whether pencil and paper or a calculator is available.
- Tending to want to "make sense" of situations involving number and quantity. Talking about numbers and their relationships.
- Having a sense of the relative size of numbers and the quantities to which numbers refer. Note that the values of numbers depend on the situational context; the same number can refer to "a lot" or "a little" in different situations.
- Substituting flexibly among different possible representations of a quantity (e.g., 24 for 2 dozen, a little less than 1/2 for 0.4). Note that a substitution's usefulness depends on the particular problem to be solved and on details of the individual's knowledge of numbers."

—Resnick, "Defining, Assessing , and Teaching Number Sense," p. 36

---

should compare different approaches and algorithms for obtaining the same results, evaluating strengths and weaknesses. They must understand why the approach they choose makes sense for the problem they are solving. If it makes sense, they will develop fluency. Making sense of a specific situation is critical.

When an explanation of a procedure escapes a student, the student probably has not investigated enough situations where the procedure might be applied. This observation also holds for other mathematical techniques, not just computational procedures. For example, tree diagrams become effective tools after students try to record results of probability experiments, explore game strategies, or design tournaments. Students should not be expected to invent tree diagrams on their own; on the other hand, teaching tree diagrams before students need them makes little sense.

Conscious choices have to be made. Depth is to be valued over pace so that the presentation of a computational procedure can be delayed until students need it and meaningful examples and motivation can be provided before the algorithm is presented. Also to be valued is the critical use of alternative algorithms. Finally, the understanding of multiplication, for example, should not be confused with fluency in using a multiplication algorithm. What is most important is to understand multiplication and to use it reliably and appropriately.

Manipulatives and Technology

The keys to appropriate use of technology in the classroom are availability, student direction, and a high-level purpose. For example, students can use manipulative materials to explore concepts rather than memorize step-by-step procedures. They can use calculators to investigate numerical patterns rather than check paper-and-pencil computations. And they can use computers to analyze data rather than perform rapid drill on basic facts. Technology in the classroom can be a positive force for equity; it helps break down barriers to mathematical understanding created by differences in computational proficiency.

Some middle school and high school students believe they are too old to use manipulatives—materials for "little kids." This resistance will persist until the students have opportunities to explore "advanced" concepts with concrete materials. Like calculators and computers, manipulatives are tools that can help make sense of mathematical concepts. The use of concrete materials and situations gives students an experiential basis for development of more abstract concepts. Concrete materials also serve as a focus of discussion in collaborative efforts, building deeper mathematical understandings and developing communication skills, even among students who do not share the same primary language.

Calculators and computers require reexamination of priorities for mathematics education. How many adults, whether store clerks or bookkeepers, still do long division (or even long multiplication) with paper and pencil? How many scientists or engineers use paper-and-pencil methods to carry out their scientific calculations?[1] Who would trust a bank that kept its records in ledger books? Electronic spreadsheets, numerical analysis packages, symbolic computer systems, and sophisticated computer graphics have become the power tools of mathematics in industry. Scientists, economists, and research mathematicians were among the first to use computers to aid in exploration, conjecture, and proof.

Technology, properly used, opens mathematical vistas not previously available. For instance, in mathematics programs that are empowering, students can use microworlds to construct and test strategies. In addition, they can use state-of-the-art software to manipulate text, graphics, and video images in ways that promote exploration and self-discovery. They can use programmable graphing calculators and programming languages such as Logo™ to illustrate important concepts and theories. Empowering programs make full use of technology without diluting in any way a student's responsibility for thinking and reasoning.

Ideally, resources in the classroom would be comparable with those in today's workplace except that the classroom would have better blocks and more beans. Students would have to choose what tools they need and when to use them and would use manipulative materials to illuminate concepts,

"In spite of the intimate intellectual link between mathematics and computing, school mathematics has responded hardly at all to curricular changes implied by the computer revolution. Curricula, texts, tests, and teaching habits—but not the students—are all products of the pre-computer age. Little could be worse for mathematics education than an environment in which schools hold students back from learning what they find natural."

—*Everybody Counts*, p. 62

---

[1]Except, of course, for those calculations made on the backs of envelopes—but that comes under "estimation."

model mathematical situations, and solve problems. They would control the technology, determining branches to take, routines to call up, or programs to write.

	Traditional practices	Some alternative practices	Recommended practices
Computational Procedures ▶	Procedures are given focus. Emphasis is placed on learning the steps to perform an algorithm and on providing enough practice so that the procedure becomes automatic for students.	Procedures are given major focus. Manipulatives are used to teach and explain the algorithm, although students are still expected to become proficient performing the algorithm with paper and pencil so that they can do well on tests that require quick and accurate computation.	Students develop a range of computational procedures, with a greater emphasis on number sense than on algorithms. They invent and use a number of different computational procedures to assist them in solving problems. Students are expected to decide the most efficient means to calculate an answer for a given situation—without assistance or with the aid of a calculator or paper and pencil.
Manipulatives ▶	Manipulatives are used in primary grades at the teacher's discretion. (They are not part of student textbooks.)	Manipulatives are used frequently but in prescribed ways to teach specific concepts or procedures.	Students have manipulatives available at all times. Sometimes students use manipulatives for specific purposes. At other times the choice of manipulatives is up to the students.
Calculators ▶	Calculators are used only for a few calculator lessons and are not permitted for daily assignments and tests.	Calculators are used for calculator lessons and problem-solving lessons. In some programs students use calculators to check paper-and-pencil computations. In other programs students use calculators after paper-and-pencil algorithms are mastered. Students are not permitted to use calculators for tests.	Students have calculators continually available for use—in class, for homework assignments, and for tests. Students conduct experiments based on using calculators to explore the behavior of mathematical structures.

Traditional practices	Some alternative practices	Recommended practices
Computers and software are used for drill and practice if at all and then often in a computer lab staffed by a specialist or aide. The computer helps the teacher only as a grade book.	Computers and software are used more frequently and more often in the classroom. Students use a few imaginative pieces of software that usually focus on specific types of problems. The computer work is considered enrichment and is not connected to students' core work. Students work in groups because there are not enough computers.	Computers are available at all times. Students can use them individually or in groups. Teachers and students use more sophisticated special-purpose programs and are fluent enough with tool software—such as word processors, spreadsheets, and data bases—to use the computer to meet their particular needs. The computer is considered to be an extension of human capability and not an end in itself.

◀ **Computers and software**

Recommendations for Calculators

Calculators are the "electronic pencils" of today's world. Mass-produced calculators are inexpensive, costing much less than students' textbooks. Even the prices of advanced calculators are reasonable and are declining. In every grade calculators can be issued to students just as textbooks are. A reasonable goal is to make calculators available at all times for in-class activities, homework, and tests.

Grade level	Type of calculator
Kindergarten through grade three	Basic four-function calculators
Grade four and beyond	Calculators with the basic four functions, plus the capacity to do powers and roots, integer division, fractions, and order of operations
Grade seven and beyond	Scientific calculators that perform routine statistical operations and probability computations
Grade nine and beyond	Programmable scientific calculators with graphing capabilities that are capable of displaying mathematical functions

Recommendations for Computers and Software

One goal is to have computers with printers and large-scale displays available in all mathematics classrooms, not only in computer labs. The mathematics program can incorporate computers and use them for both group and individual work. Using the computer primarily for drill and practice is inconsistent with this *Framework*. The use of software that places the student in a more constructive role is a richer environment for developing mathematical power.

"Learner-centered software puts students in control of learning activities and challenges them to generate and solve problems as diverse as writing a story, solving a mystery, or creating an animated picture on the computer screen. The child becomes actively involved in charting the learning course that he [or she] will follow while engaged with the software. Learner-centered software encourages the child to think and to use information to solve problems—or to create new problems to solve.

Learner-centered software is characterized by four important features:

1. Learner-centered software offers students choices in selecting the goal of an activity, the strategies to reach the goal, or both.
2. Learner-centered software provides feedback that is informational rather than judgmental.
3. Learner-centered software allows, emphasizes, or encourages prediction and successive approximation.
4. Learner-centered software encourages learning within a meaningful context for students, building on students' intrinsic motivation."

—Russell and others, *Beyond Drill and Practice: Expanding the Computer Mainstream*, pp. 3–6

New technology-based materials exceed the capacity of hardware purchased in the late 1970s and early 1980s. The materials can be used to augment current hardware for the 1990s. Students need more interactive hardware and software to explore patterns, manipulate data, and design graphics. To use other hardware systems, including video disks, scanners, laser printers, plotters, modems, and complex networks, students will probably need computers with 32-bit processors, bit-mapped displays, and hard disks.

Grouping Students

This *Framework* calls for heterogeneous grouping and untracking as a goal and recommends against attempting to group students by ability. The time students spend in effective heterogeneous groups can be increased significantly—and time spent on lecture and seatwork decreased—through a practical, balanced approach involving a mixture of strategies and cautions:

1. Students who need more time to sustain success are given that opportunity before and after school, on Saturdays, and during the summer session or intersession.
2. At times the class might be broken into temporary groups with similar interests or needs; that is, students who:

   a. Have gaps in their understanding of the mathematical ideas needed to participate productively in core activities.
   b. Need to develop the techniques and tools they will use in the core activity.
   c. Are interested in going deeper or further or working on a class presentation or an ambitious project together.
   d. Share a language other than English.[2]

3. Supplemental instruction does not pull students out of the core curriculum in mathematics or any other subject.
4. Instruction is provided in collaboration skills for effective work in groups as part of the mathematics curriculum.
5. Students are not urged to race ahead. Racing through a curriculum that bores them does not enhance learning. Instead, students are encouraged to probe more deeply or extend their efforts to related areas. In this way they will experience a more interesting and challenging engagement with mathematics.[3]
6. Students are not transferred out of the core into a watered-down curriculum. Facility with skills is *not* a prerequisite for participation in the core curriculum; in fact, it should be developed in conjunction with core work.
7. Students are not sorted into groups by perceived ability because this practice creates a self-fulfilling prophecy of low performance. In elementary school this practice amounts to tracking.

Avoiding Ability Grouping

Ability grouping takes different forms in different grades. Elementary classes are sometimes grouped by reading level or mathematics ability; middle school students get tracked into

---

[2]But see the section on students who are developing English proficiency, p. 45.
[3]See comments on gifted students, pp. 65–66.

different classes of the same course; and high school students find themselves in different course sequences.

Even if it is believed that there are large, significant differences among students in their capacity to learn mathematics, there are problems with ability grouping for all ages. Research indicates that long-term grouping according to ability often harms students in the lower and middle groups and rarely helps those in the higher groups. Once students are placed in a low group, often in a sincere attempt to accommodate individual differences, they tend to stay in it during their whole school career and come to see themselves and are seen by peers and teachers as less competent than other students.

Instead of helping students, sorting and tracking them according to ability institutionalize failure in mathematics. However, placing students in heterogeneous classes and groups and teaching the same old curriculum will not solve the problem. Coping with students who differ in maturity and mathematical development requires curricular reform. The curriculum must be untracked just as the school structure must be untracked. A multidimensional curriculum will be accessible to more students and more interesting and valuable to the most mathematically sophisticated. When schools focus on the depth and quality of thoughtful work, they will make heterogeneous groups and classes more practical for students and teachers alike.

Many teachers have not been exposed to or do not have access to curricula appropriate for heterogeneous classes. Teachers need both curricula and time to plan before untracked classes can be made to work well for all. The move toward untracked mathematics will require patience, cooperation, and hard work.

Moving Away from Tracking

Tracking is a word that triggers strong feelings, often hostile, among proponents and critics. However, a climate can be fostered in which both camps exhibit patience and problem-solving skills in formulating their opinions. In fact, it is time to move beyond taking stands and to start taking steps to solve the real problem—that at each grade level too many students stop growing mathematically. The system fails too many students each year.

How can a mathematics program work for students with a wide range of sophistication and interest in mathematics? The accompanying questions on tracking might help educators develop a comprehensive strategy. Everyone—teachers, parents, and students—should think about these issues and discuss them.

Tracking: Questions to Consider

1. Are decisions about placement made in the interest of lifting students to the most challenging course they can take?

2. Are all students expected to do complete work by thinking, communicating, drawing on mathematical ideas, and using tools and techniques?

3. What services are available to help students with gaps in their skills and understandings to participate successfully in core mathematics programs?

4. Can students with special interests in mathematics pursue mathematical issues with similarly interested peers while participating in core programs?

5. What help is there for students doing homework? Is there a time and place for students to work together or alone on homework?

6. Do students have the chance to revise and resubmit their work until it meets quality standards?

7. Do teachers show the same enthusiasm for teaching different classes? Are some classes considered plums? Are others low or troublesome?

8. Are students from the various ethnic and socioeconomic groups represented in the school distributed equitably across the curriculum?

9. Do the courses ensure that all students will be able to maximize their mathematical education; i.e., not inhibit some students from studying advanced mathematics and restrict others to dead-end tracks?

10. Do the teaching methods in all classes accommodate cultural diversity and varied learning styles?

11. Do all courses promote the development of mathematical self-esteem?

12. Do all students have the opportunity to study all of the mathematics courses offered?

13. Do all students have access to the same quality of technology?

14. How are students assessed? Does the program assess mathematical thinking and understanding? Are tests biased against any ethnic or socioeconomic group?

15. Do students and parents participate in decisions about placement?

The situation in eighth grade is especially difficult. Should some eighth graders begin the high school mathematics sequence? Many eighth-grade courses repeat the curriculum offered earlier. In these circumstances an early start on the new material in high school makes sense. As the eighth-grade and ninth-grade courses improve, the situation is expected to change. Eighth graders, even many who may become scientists or engineers, will benefit more from a worthwhile eighth-grade course than they would from early entry into the high school course of study. However, some eighth graders may want two full years of high school mathematics beyond the three-year core sequence (see Chapter 5) and should be able to begin high school mathematics early.

At the other end of the spectrum, how can all eighth graders be ready for a richer and more thoughtful high school curriculum? One possibility is a summer or intersession program for some middle grades students. The focus of this program should be on getting a head start on the big ideas and difficult mathematical issues contained in the high school curriculum. Even though there may be gaps in these students' understanding from the elementary school and middle school curricula, it is better to look ahead than to remediate.

In high school the mathematics curriculum has been differentiated by grade level into two sets of tracks, a college preparatory sequence and a computational proficiency sequence. The former serves as the primary filter, determining who becomes eligible for entrance to four-year institutions. Algebra eliminates the largest number. Many students who receive a C or D grade in the course despair of any future mathematical work long before they finish high school, even if they pass geometry and Algebra II. The picture is worse in the lower tracks, where students cycle repeatedly through fractions and percents, are barred from thinking, and are often disenfranchised and labeled as unteachable. This filter does not distribute misery evenly across the student population. African Americans and Latinos are underrepresented in higher education in general and in mathematically based programs of study in particular.[4]

Tracking cannot be eliminated in a short period of time or by edict. Nevertheless, on the basis of the experiences of educators

"... Studies that ... provide identical curriculum and instruction to both tracked and mixed groups of students typically find that high-ability students do equally well in either setting."

—Oakes, "Tracking: Can Schools Take a Different Route?" 43

---

[4]The gap for female students has narrowed greatly in the last generation but is still significant. The difference in expectations for females in kindergarten through grade twelve is subtle and hard to document, and the statistical gap is negligible until grade twelve, when any student who wishes to pursue mathematically based programs should be in an advanced mathematics class.

testing new, untracked models for mathematics courses ranging from grade seven through grade eleven, this *Framework* recommends the elimination of all general mathematics, consumer mathematics, and foundations courses in grades seven through twelve. These shallow courses, which emphasize repetition and low-level skills, must be replaced with new, well-designed, substantial courses that are suitable for all students and focus on depth and important mathematical ideas. (More information on how such courses might be structured is presented in Chapter 5.)

For *mathematically gifted students*, emphasis should be placed on the quality, depth, and originality of their work rather than on speed and accuracy. Like other students, mathematically gifted students need to explore ideas in depth and think about a variety of approaches and perceptions. They need time to work with other students who are similarly interested in going further or deeper into mathematical issues. Paying attention to such special interest groups should be a regular feature of heterogeneous classrooms. Instead of emphasis on acceleration through the core curriculum, the breadth and range of math-

"...Precocious students get the idea that the reward is in being 'ahead' of others in the same age group, rather than in the quality of learning and thinking. With a lifetime to learn, this is a shortsighted attitude. By the time they are twenty-five or thirty, they are judged not by precociousness, but on the quality of work."

—Thurston, "Mathematical Education," 846

Traditional practices	Some alternative practices	Recommended practices	
Students are put in ability-level groups (elementary) or tracks (secondary), mainly on the basis of how quickly they can perform paper-and-pencil computations or algebraic manipulations. In elementary school, students often work in the same class and study the same material but at different rates. In secondary school, students are grouped in separate classes with different titles and study vastly different materials.	In elementary school the teacher teaches the heterogeneous class as a whole group but pulls out and gives additional assistance to students who are having difficulties. Some secondary schools institute Math A classes. Secondary teachers discuss tracking problems but see no workable alternative to the traditional college preparatory courses.	All students study a common core curriculum. In kindergarten through grade eight, classes are grouped heterogeneously. Students with special interest or talents go more deeply into some investigations, often working with an interest group. Students having difficulty sometimes get additional assistance, often before or after school. Some students begin the high school core sequence early, most in ninth grade, and the remainder after completing Math A and, perhaps, Math B.	◀ **Grouping**

ematics should be exploited in the design of enrichment materials for especially talented students.

Unusually talented young mathematicians need even more experience working with a cross section of their age group. When enrolled in classes for the gifted or pull-out programs, they are deprived of important perspectives and learning opportunities for working on complex tasks that they would enjoy if enrolled with a more representative cross-section of their peers.

Assessing Student Work for Mathematical Power

In the past decade an explosion of new thinking about assessment in mathematics has occurred. As with collaboration in the mathematics classroom, a long description of current thinking on assessment is beyond the scope of this document and would duplicate the excellent work of others. Readers may refer to the following resources for more extensive information:

Resources on Assessment

In addition to documents liberally cited elsewhere in this publication, such as *Everybody Counts* and the NCTM *Standards*, the following are excellent resources for learning what people are doing with innovative assessment in mathematics:

Educational Leadership, Vol. 46, No. 7 (April, 1989). This issue is devoted entirely to assessment.

Mathematics Assessment: Myths, Models, Good Questions, and Practical Suggestions. Edited by J. K. Stenmark. Reston, Va.: National Council of Teachers of Mathematics, 1991.

Mumme, J. *Portfolio Assessment in Mathematics.* Santa Barbara: Tri-County Mathematics Project, University of California, 1991.

A Sampler of Mathematics Assessment. Sacramento: California Department of Education, 1991.

Stenmark, J. K. *Assessment Alternatives in Mathematics.* Berkeley: Lawrence Hall of Science, University of California, 1989.

The type of assessment recommended in this *Framework* is often called *authentic* assessment to signify that it looks at what students do and know rather than at some proxy. Sometimes the phrase *performance assessment* is used to help focus on student work and action rather than on the students themselves.

The authenticity of this assessment has some important implications:

1. Assessment takes place in the course of normal work. The class and the learning do not stop for a test. The work and the assessment are inseparable.
2. The conditions for assessment, like the conditions for meaningful tasks, mirror the conditions for doing mathematics outside the school. That is, students have ample time; have access to colleagues and to tools (books, notes, calculators, pattern blocks, and so on); and have the opportunity to revise their work.
3. The tasks for assessment, like all meaningful tasks, engage a student's sense of purpose and are rich enough to be multidimensional. They allow students to demonstrate thinking, understanding, and communication skills as well mastery of techniques. These tasks may result in unexpected but valid responses.

In addition, this kind of assessment is respectful of students:

1. Feedback from assessment informs students' efforts; it is concrete and specific to the tasks.
2. Students participate in the process of assessment; they help to create and apply standards for quality work. Self-assessment and peer assessment help ensure that assessment schemes are acceptable and comprehensible to students.

Traditionally, the purpose of assessment was simply to evaluate students, to check on their skills and abilities. A broader view of assessment encompasses more varied purposes:

1. Assessment provides *teachers* with a window into students' thinking and understanding, as well as into their proficiency with more narrow skills, and reveals the quality of the students' mathematical communication. It also helps teachers evaluate the success of the mathematics program and provides information helpful in making decisions on instruction.
2. Assessment can show *administrators*, *parents*, and the *public* a more comprehensible, three-dimensional picture of

"Increasingly we find advantages in blurring the lines between instruction and assessment, in drawing assessment information from instructional tasks, and in providing additional instruction in the context of assessment activities."

—*Mathematics Assessment*, p. 5

67

children's learning and progress. It also helps everyone see what the program is like and what is expected of students. Other adults can help only if they understand the program.

3. Assessment can give *students* what they need to improve the quality of their work. Of course, progress comes through revision based on the advice that assessment provides; but it also comes through students' participation in the assessment process. Good assessment facilitates learning and independent thinking, partly because it helps the students themselves to recognize effective work rather than rely on the judgment of others.

Alternatives to Traditional Assessment

Various assessment techniques are emerging, including ways of assessing student work on open-ended tasks, observing students at work, and keeping student portfolios. These techniques complement one another and provide different views of a student's mathematical sophistication. The use of different approaches together helps "triangulate" a student's performance. As a result, some teachers have found that they can already eliminate most fill-in-the-blanks tests. Three alternative sources of assessment information are the following:

1. *Open-ended tasks.* These tasks can take from ten minutes to many weeks. Together, teachers and students can learn to evaluate work on this kind of task. They can develop their expertise in designing rubrics and using them to evaluate shorter open-ended responses holistically.[5] Investigations produced over a period of weeks with drafts and revision can be looked at in the same way. Scoring rubrics may be generalized or specific to the task, like the one shown in Chapter 1 (page 29). If students learn to score and improve their skill in evaluating their work, the quality of the work will improve. Involving students in reviewing each other's drafts also increases feedback without increasing teacher workload. Teachers and students can return periodically to evaluate their assessments to make sure that everyone understands the standards of quality in the same way.

---

[5] If open-ended items must be scored for external evaluation or comparison, a group of teachers might allow time for "calibration" to determine a common standard for the scores. If this experience is new to a school, a workshop on scoring student work generated in the California Assessment Program process might be offered.

2. *Observations of students at work.* Afterwards, the observations can be discussed with the students and with other teachers. The students may be the teacher's own or students from another class, working alone or in groups. The observations may be silent, or the teacher may interact with the students with a preplanned set of questions (a protocol) or spontaneously. The teacher should make some notes during the observation period for later discussion.

3. *Student portfolios.* There is no standard way to structure a portfolio, which may include a wide range of work to address the variety of purposes or audiences intended for the student's work. A well-designed portfolio can be a robust indicator of a student's mathematical power and reveals much about the effectiveness of different assignments. A monthly review of portfolios also provides the faculty with effective feedback for program modifications.

In any of these (and other valid) approaches, a key point is that assessment is integrated with normal instruction. The purpose of the writing or presenting is to enhance the student's clarity of thinking and ability to communicate. The assessment opportunity is a by-product.

Students deserve to know what is expected of them. Performance standards for portfolios, for example, are subtle. Proce-

Writing and Elegance

Writing clarifies and reveals thinking. Many practitioners have found student writing to be the most important ingredient in changing the way they assess performance. Teachers of writing are an important resource for teachers of mathematics; they know how to help students write well and to help one another. Even first and second graders can write about mathematics. Here are two issues we need to address:

First, students do not yet expect to write in mathematics class. It may be strange to them; they often do not know how to transfer what they do in language arts. Some students like mathematics precisely because they thought they would not have to write. Students with developing English writing skills are at a disadvantage if written products are graded on the basis of written fluency. Instead, teachers should evaluate written work in mathematics on the basis of the thinking it reveals. Students who are learning English often write more effectively in mathematical situations than they do in language arts, and many "gifted" students demonstrate problems with communication that go unnoticed on bubble tests.

Second, we must help students strive for elegance in their communication—to learn to explain clearly and concisely. It is tempting to award high marks to a neat paper with pretty pictures and lots of well-formed sentences—but little mathematics. We must concentrate on the mathematical power student writing reveals rather than on the writing itself.

dures and standards must be honest and clear. An excellent approach here is to have students participate, letting them help define criteria for good work and then evaluating student work (their own or that of colleagues) in comparison with those criteria. Another way to show students what is expected is to model good work, letting students see the teacher creating first-rate responses to open-ended tasks.

Tests and assessment ▶

Traditional practices	Some alternative practices	Recommended practices
Testing is important. The program may use the test-teach-test cycle. Tests closely match assignments, with a narrow focus on skill. Questions have one correct answer. Tests can be scored objectively. Student mastery is expected. The goal of testing is to assess—and classify—the student.	Formal tests are important. Although teachers question their usefulness and their impact, they expect their students to do well on norm-referenced tests. Some teachers include open-ended questions and require students to explain their thinking on their tests.	The goal of assessment is to evaluate student work rather than the student. There are many forms of external and internal assessment, including portfolios of student work, observation, interviews, and group work. Students are expected to explain their thinking. Holistic scoring rubrics focus on all dimensions of mathematical power. Often, students are expected to revise their work to meet the standards. Students participate in assessment.

Reporting of Student Progress

Assigning a single letter grade or numerical grade to a student's work or test rarely communicates the complex, multi-dimensional aspects of a student's evolving mathematical accomplishments. Grades sometimes become more important than what the student learns. Sometimes grades tell students only that they are "not good at math." Formal assessment and feedback should encourage, not inhibit, the development of mathematical power. The key to success is to evaluate work and performance according to well-defined standards and not sort out and stigmatize students.

Parents want to know how their child is doing in the study of mathematics. In addition to grades, or instead of them, teachers might describe, orally or in writing, a student's mathematical work and the progress shown. Teachers might show parents the portfolios and mathematics journals (created by students) as evidence of student development over time. At parent conferences the teacher might have the child do some mathematical task. Parent conferences may present less an opportunity to report than to discuss a student's mathematical power, with the student participating in the discussion.

Developing a Positive Disposition Toward Mathematics

How do some persons develop the idea that mathematics is something that they are good at and find worth doing, others see it as a chore, and still others give up before they have completed the minimum for educational or employment options? Many Californians think that mathematical talent is something people are born with. Yet students from other countries who outscore American students in mathematics think that it is effort rather than native ability that leads to success. Students will be more willing to persist when it is acknowledged that mathematics is challenging and when they understand that the goal of mathematics is to make difficult problems solvable.

A second contributor to students' judgments about their mathematical abilities is the selection of problems to work on. Very difficult puzzles that some students solve quickly but baffle everyone else may help identify quick puzzle solvers but do not teach those students much mathematics. Even worse, they teach everyone else that mathematics is not accessible to them. Some of the best problems absorb a week's worth of attention and require thinking with knowledge and persistence, not cleverness. For example, a good assignment might be the following:

What is *chance*? Explain your ideas about chance. Give examples of chance in three or four different situations, interpreting what happens according to your ideas of chance. Use diagrams, numbers, algebraic symbols, and other mathematical techniques to communicate your ideas.

This assignment is accessible to a wide range of students, especially if they are given the opportunity to revise their work on the basis of feedback. Students will respond to the assignment differently according to their maturity and level of mathematical development; but each one will learn something about a big idea in mathematics, and each will have tangible evidence of accomplishment. Assignments like this might appear before and after a unit—on fairness or expected values, for example—and be used to assess student growth.

A third factor in developing students' positive attitudes toward doing mathematics is the way in which they are allowed to work on problems or investigations. Are a variety of approaches considered valuable? Do students have opportunities for discussion with peers and consultation with the teacher? Are tools available? Are students given feedback and the opportunity to revise their work? Is there a cooperative, problem-solving atmosphere that promotes healthy confusion, honest disagreement, and playful use of imagination?

Other behaviors help form a productive attitude toward mathematics. Students should look for problems as well as solve them. Students will naturally stumble upon other questions and often ask *what if?* while working on a problem. They can be encouraged to pursue those questions and formulate new problems or conjectures. In essence students are doing what mathematicians do; they are "looking for trouble"—the attitude that is at the heart of research.

Special Considerations for the Primary Grades

The developmental levels of all students must be considered when the program recommendations in this *Framework* are being interpreted. Very young children view the world differently from older children and adults. Thoughts are more dependent on their senses—what they see, hear, or touch—rather than on more abstract knowledge true to a given situation. For example, young children:

1. Have a hard time holding more than one idea in their mind at a time. They may not grasp that *some* of the blocks in a pile are red and at the same time *all* of the blocks are made of wood.
2. Have shorter attention spans, especially if a task has no meaning for them.
3. Are more egocentric. They have a difficult time understanding another's perspective.
4. Often work on a task in parallel alongside other children. The children work individually, talking about what they are doing but not collaborating.
5. May be able to replicate practiced routines but will not have the same understanding of the tasks that an adult will have. Adults may observe a young child performing what appears to be a multistep procedure and wrongly infer that the child has developed the complex understandings that underlie the task.

Because children's ideas of who they are and what they do are equivalent, grouping children by ability in the primary grades can be especially damaging because of its harmful effect on self-image. Emphasis at this age should be placed on accomplishments rather than shortcomings and on participation rather than achievement.

The teacher should provide meaningful learning opportunities and then observe, support, and guide the children along the way because young children learn best from concrete experiences and meaningful play. They will assimilate what makes sense and what is useful.

Although primary grade students should be assigned tasks and investigations that are relatively short in duration and are divided into smaller parts, they work successfully with problems that require more time. For example, if the students are planning how to spend a P.T.A. cash gift for class materials, the teacher may structure the activity so that the students determine what the class wants and needs on one day, review the prices of the new items on another day, and then check to see how their choices match their needs and budget on a third day.

Students respond best to repeated encounters with mathematical ideas cast in different contexts in which they talk or write about each experience. This strategy helps students make connections among ideas. Young students should not be expected to generalize from their experiences.

Acquisition of Number Facts

Elementary students' mathematical thinking is concrete, informal, and playful. Students can show their thinking in a variety of ways—by using concrete materials, talking with peers, drawing, and writing.

Students with single-digit number facts at their fingertips can perform common computations quickly and easily. Without this ability students can be stunted in developing number sense. However, the emphasis on *early* memorization of basic facts has hindered rather than helped mathematical understanding. For students who do not memorize basic facts quickly or easily, mathematics can become drudgery. They can believe that they are not as competent as other students, and these feelings of inadequacy can affect their ability to memorize. To help students overcome this problem, teachers may have to postpone emphasis on facility with basic facts.

Only after students understand the meanings and uses of the operations will they see how the basic facts will help them solve problems more efficiently. Many children will learn the easier facts quite naturally in the context of doing mathematics so that explicit practice on them will not be necessary.

For other children the teacher can be a facilitator and supporter, helping them see patterns and strategies for memorizing the facts. Helping students see the benefits of memorization for themselves convinces them that memorization is worthwhile. Demonstrating that there are only about 20 multiplication facts to memorize makes the task seem manageable and perhaps can be accomplished in a month if only one fact is learned each day.[6]

Rather than checking off each fact as a child memorizes it, teachers can tell which students know the basic facts by observing the students at work. Teachers may ask: Are students counting on their fingers or making tallies while multiplying? Are they becoming frustrated because they have to figure out a basic fact every time they need it? The teacher's support of students as they look for patterns and strategies to learn the basic facts will help relieve the stress many students feel.

---

[6]Many students think of the multiplication table as 100 distinct facts and fail to realize that what they understand about multiplication makes their tasks easier. Understanding the commutative property (e.g., that 3×6 is the same as 6×3) cuts the number of facts to memorize in half. And understanding what multiplying by zero and one means eliminates the need to practice these facts.

Structure and Content of the Mathematics Program

C h a p t e r **3**

This chapter is concerned with the overall structure and content of the mathematics program. First, two different ways to look at mathematical content are discussed and related to the strands of the 1985 *Framework*. Next, units of instruction, the structural building blocks of the program, and their relation to this *Framework's* view of content are described. Also discussed is the selection of units to make up a year's work. The content for different grade levels will be discussed in Chapter 4 and Chapter 5. (See Appendix A for the criteria that should serve as standards for the statewide adoption of mathematics instructional materials, kindergarten through grade eight, in 1994.)

Although many of the recommendations made in Chapter 2 can be put directly into practice by teachers and administrators, the same cannot be said of the recommendations made in this chapter. Implementation will require that curriculum developers begin with the design of new programs and materials. Although this chapter will be useful to teachers when they select materials, design lessons, and make day-to-day instructional decisions, it is directed primarily to those who develop curriculum materials as their principal contribution to education.

A New Curriculum

As discussed in Chapter 2, much can be done to promote mathematical power without creating new instructional materials. For example, classrooms can be organized to improve communication; students can be helped to learn to write about mathematics and can be allowed to talk and work with one another; technology can be made available; and students can be helped to develop a positive disposition. These things alone, however, will not make students mathematically powerful. The missing ingredients are the mathematics content and a structure in which to present it. In terms of mathematical power, students need contexts in which to think and communicate. They need tasks for which mathematical tools and techniques are appropriate and that embody the mathematical content students are expected to learn.

Large mathematical ideas and their interconnections are at the heart of the program. Students do not become mathematically powerful by mastering isolated bits of mathematics. Traditionally, students were expected to become powerful once the ingredients were all covered, but they did not attain that goal. Instead, teachers need to depend on the students' desire to make sense of interesting situations by using mathematics that helps the students to do so.

To create opportunities for students to build their understanding, the program must provide access to the mathematics through inherently interesting contexts and diverse experiences. At the same time the content must relate to other content so that students can see and use mathematical connections to make better sense of the world. The mathematical ideas must then be large enough to make interconnections possible and rich contexts believable.

This *Framework* proposes that these rich contexts, together with coherent bodies of mathematics and integrated assessment, appear in instructional *units* that contain large and interconnected assignments, such as investigations. Students also need ample experience with the varied tools, techniques, and skills of mathematics. Much of this experience will arise as they work on larger problems and projects; but straightforward showing, doing, and explaining of mathematics are also needed. The section in Chapter 1 (pages 17–19) describing exercises, problems, and investigations is useful in getting a feel for the mixture of the tasks students might do.

Units give students a balanced mixture of assignments organized around large tasks. Before these units of instruction can be created, however, mathematical content must be described in a way that facilitates interconnections and promotes large pieces of work. This *Framework* next specifies mathematical content, with an eye toward understanding how the mathematics will appear in units of instruction.

Specification of Mathematical Content

This *Framework* bases its specification of mathematical content squarely on the *Curriculum and Evaluation Standards for School Mathematics*, published by the National Council of Teachers of Mathematics (NCTM) in 1989. At the same time it builds on these standards by describing content through a two-part perspective of *strands* and *unifying ideas*. After a brief characterization of the *Standards*, the strands and unifying ideas will be discussed.

NCTM *Standards*

The *Curriculum and Evaluation Standards* presents a set of standards for evaluating the quality of both the mathematics curriculum and student achievement in that curriculum. The document represents the consensus of NCTM's members about the fundamental content that should be included in the school mathematics curriculum and key issues concerning student and program evaluation.

At the heart of this consensus are the basic tenets that (1) all students need to learn more (and often different) mathematical content than is contained in current programs; and (2) the currently prevalent methods of instruction in mathematics must be significantly revised. This document is intended to establish a broad framework based on these tenets to guide reform in school mathematics over the next decade. It gives a vision of what the mathematics curriculum should include in terms of priority and emphasis in the content.

The *Curriculum and Evaluation Standards* identifies the term *standard* as a statement that can be used to judge the quality of a mathematics curriculum or of methods of evaluation. A standard is meant to be a statement about what is valued.

In keeping with the national consensus on mathematics education, this *Framework* endorses the NCTM *Standards* and incorporates those standards as an expression of the desired outcomes of the curriculum for kindergarten through grade twelve.

Perspectives on Content Specification

The content for mathematics programs is usually specified in a listing of topics for each course. Lists have obvious value; they tell authors what content to include in a given course, and the lists let teachers of subsequent courses know what has already been covered. However, describing content through lists has some disadvantages. Lists typically do not emphasize major ideas, distinguish key concepts from secondary detail, or identify unifying principles that tie together different topics. Nor do they convey the setting or the spirit in which the mathematics is to be approached. For these reasons the course content is described differently in this *Framework*, with major ideas and connections attended to. Although lists are not abandoned completely, their reduction in number allows more emphasis on global features

"Traditional school mathematics picks very few strands (e.g., arithmetic, geometry, algebra) and arranges them horizontally to form the curriculum: first arithmetic, then simple algebra, then geometry, then more algebra, and finally—as if it were the epitome of mathematical knowledge—calculus. This layer-cake approach to mathematics education . . . reinforces the tendency to design each course primarily to meet the prerequisites of the next course, making the study of mathematics largely an exercise in delayed gratification. To help students see clearly their own mathematical futures, we need to construct curricula with greater vertical continuity, to connect the roots of mathematics to the branches of mathematics in the educational experience of children."

On the Shoulders of Giants,
p. 4

and questions of relative importance. In addition, this approach follows the pattern adhered to in other frameworks.

No single way of looking at the mathematical content of a program is fully adequate. Therefore, content is described from two different perspectives, *strands* and *unifying ideas*. Each characterizes content from a particular viewpoint. Together, these two perspectives give a more balanced picture than a single perspective, describing the specific mathematical content of the program and corresponding to the mathematical ideas dimension of mathematical power.

The relationship between the standards of the *Curriculum and Evaluation Standards* and the *strands* and *unifying ideas* of this *Framework* must be clearly understood. Although the *Framework* endorses fully the message of the *Standards*, the nature of the classification scheme represented by standards differs somewhat from the scheme represented by the strands and unifying ideas, which are described in the next part of this chapter and in chapters 4 and 5.

One difference between standards and the strands and unifying ideas is that the standards represent a broader perspective. The first four standards for the kindergarten through grade twelve curriculum (dealing with problem solving, communication, reasoning, and connections) involve the *process* of doing mathematics, and standards 5 and up are *content* standards. Issues of process are of extreme importance in this *Framework* and receive separate and full discussion in Chapter 1 under *thinking* and *communication*, which are two of the four major dimensions of mathematical power. Strands and unifying ideas represent a way of talking about content alone and belong under a third dimension of mathematical power, *mathematical ideas*. This third dimension is developed in this chapter. Another difference is that all the standards, including the content standards, deal explicitly with a wider range of issues than do strands and unifying ideas; for example, with the quality of students' achievement and with methods of mathematics instruction.

Further, the content standards represent three different slices in time for kindergarten through grade four, grades five through eight, and grades nine through twelve. The names of the standards and their particular content show some relevant differences from level to level. Strand categories, on the other

hand, do not represent a level-specific classification. They are continuous threads running throughout the curriculum, each developed in appropriate ways at all grade levels, kindergarten through grade twelve. (Clearly, appropriateness is the key here. Just as only an appropriately simple conception of algebra should appear as early as kindergarten through grade three, only an appropriately sophisticated conception of number should claim attention in grades nine through twelve.)

Finally, the standards represent a somewhat more fine-grained classification of content than do strands. For example, several standards correspond to the number strand in kindergarten through grade four. And there are standards in grades nine through twelve, such as "Geometry from an Algebraic Perspective" and "Trigonometry," that correspond to merging of ideas from different strands.

Despite these differences there is a large degree of correspondence between the content standards (numbered 5 and up) and the strands of this *Framework*. The descriptions of particular strands in this *Framework* need to be read in conjunction with the relevant standards.

A way in which the present *Framework* steps beyond the *Standards* is through the use of unifying ideas, which represent mathematical content issues that are relevant in several different strands (equivalently, in several different standards). They respond to the stated intention in the *Standards* that the standards in practice need to be "integrated across courses" (page 136).

Not everything important in mathematics education is strictly content. Other important program features, such as work with problem solving and the use of manipulatives and computers, are not content. They are treated in Chapter 1 and Chapter 2.

Strands

The strands for mathematics in kindergarten through grade twelve are the traditional, widely used subject categories of mathematics. Strands have appeared in California mathematics frameworks for about 30 years.

Strands provide a referent for assessing the balance of content; they help broaden the scope of school mathematics. It is not acceptable for elementary school mathematics to be concentrated solely on arithmetic or for high school mathematics to be concentrated solely on algebra and geometry. Appropriate material from each strand should appear at each grade level. Together, the strands describe the range of mathematics important for today's students.

Yet it is not enough simply to represent all the different strands somewhere in the curriculum. The strands are meant to *interweave* or *integrate* with one another. Some content in fact involves particular mathematical principles that belong equally to several different strands. These principles will be discussed under *unifying ideas*.

Real mathematical problems rarely involve just one strand. Rather, they demand that the problem solver integrate ideas from several strands to arrive at a meaningful result. Therefore, every unit of mathematical work must involve an appropriate mixture of ideas from more than one strand. Each of the strands is described as follows:

Functions

The *patterns and functions* strand has been replaced simply by a *functions* strand because the role of patterns in mathematics is too important and too broad to link with one particular strand such as functions. In fact, patterns play a prominent role in all strands of mathematics: *functions* (functions often represent a way of generalizing a numerical pattern); *geometry* (e.g., the role of patterns in tessellations); *number* (e.g., the Fibonacci sequence); *algebra* (e.g., the binomial theorem); and *statistics* (where a main goal is to find patterns in the data). Because it crosses many strands, the theme of *patterns* is an example of what this *Framework* calls a *unifying idea*. Unifying ideas are another way to analyze curriculum content; they focus attention on ideas that apply to many strands. Unifying ideas are discussed later.

Functions remains as a strand in this *Framework* because it represents important and useful mathematics. By labeling the strand *functions*, however, the formal mathematical definition of function (a mapping of all members of one set to members of another) is not being emphasized. Instead, the functions strand explores quite broadly the many kinds of relationships among quantities and the manner in which those relationships can be made explicit—but not necessarily symbolic.

In the early grades functions almost always appear in conjunction with another strand. For example, when primary students explore the number of cookies they need for different numbers of friends, they are exploring functions as well as number. And when elementary students learn to find the area of a rectangle, they are studying functions as well as geometry. Conversion formulas such as $F = 32 + 9/5\ C$ and distributions of die rolls are functions as well. Students making polygons with Logo ™ discover how the angle at which the turtle turns depends on the number of sides.

Functions are a backbone of mathematics but need not be the formal, abstract, imposing, and opaque endeavors that traditionally terrorize Algebra II students. Thoughtful experiences with a variety of functions, formal and informal, in early grades will help all students make functions part of their intellectual repertoire when the functions are met later on.

2 Algebra

Algebra is generalized arithmetic. It helps make the specific universal, describe situations, and derive relationships with elegance and power. By itself it is the language of variables, operations, and symbol manipulation. In addition, every other strand uses algebra to symbolize, clarify, and communicate. It is important, therefore, to use algebra in the context of problems and situations arising in other strands.

The relationship between the *algebra* and *functions* strands deserves particular attention. Algebra helps in the manipulation of expressions and the finding of solutions; functions describe relationships among quantities. The point emphasized previously—that strands must be integrated with one another—applies here especially. These two subjects remain far too separate in most treatments of school mathematics.

Specifically, teaching formal algebra without appropriate ideas about functions makes the algebraic ideas more mysterious, harder to grasp, and harder to link with future work. (Some high school students miss the fact, for example, that polynomials are functions and that the essence of solving equations involves finding points where two functions have common values.) At the same time an understanding of functions (when expressed symbolically) requires algebra from the very start to provide the *variable expressions* that describe functions.

For similar reasons the curriculum needs much more of an integration of *algebra* and *geometry* to replace the typical separation of these subjects at the high school level into year-long, single-strand segments. Specifically, students need the tools of algebra (such as formulas, functions, and equations) to describe and clarify geometric relationships; and they need the vehicle of geometry to provide graphic illustrations of algebraic relationships.

In the early grades students should not be expected to use symbols to represent variables and solve equations on paper. Instead, they can be expected to develop the concept of a variable and of algebraic operations informally. In solving some problems with manipulatives, for example ("Take a block and let that be the number of beans Jack has in his bag"), students use concrete objects as symbols they can manipulate. The block becomes a variable years before x appears. Missing addend problems help students develop intuition not just about number but about what must occur around an equals sign in order to arrive at a solution. As students' work in all the strands of mathematics becomes more sophisticated, their knowledge of and ability to use algebra as a language should develop correspondingly. Older students use algebra to express symbolic relationships, model interactions between quantities, and extract critical information about these relationships.

3. Geometry

Through the study of geometry, students link mathematics to space and form in the world around them and in the abstract. In this strand students are exposed to and investigate two-dimensional and three-dimensional space by exploring shape, area, and volume; studying lines, angles, points, and surfaces; and engaging in other visual and concrete experiences. In the early grades this process is informal and highly experiential; students explore many objects and discover and discuss the attributes of different shapes and figures.

Older students gradually build on their foundation of hands-on experience. They become more familiar with the properties of geometrical figures and get better at using them to solve problems. They explore symmetry and proportion and begin to relate geometry to other areas of mathematics—to the benefit of both. For example, graphical representations of functions can help explain and generalize geometric relationships while geometrical insights inform the study of functions.

As students become more familiar with geometrical figures, they are better equipped for mathematical argumentation in that field. They focus on making convincing arguments with a rigor appropriate to the situation rather than on being forced into two-column proofs. The goal is to develop fluency with basic geometrical objects and relationships and to connect that fluency with spatial reasoning and visualization skills.

Statistics and Probability

In an age of rapid communication and immediate access to information, data abound. Descriptive statistics help students learn to collect and organize information in a variety of graphs, charts, and tables to make those data easier for the students and others to comprehend. Students will also learn to interpret data and to make decisions based on their interpretations. Probability is a part of this strand because statistical data are often used to predict the likelihood of future events and outcomes. Students learn probability, the study of chance, so that numerical data can be used to predict future events as well as record the past. A command of statistics and probability is essential in all aspects of adult life.

Discrete Mathematics

The *discrete mathematics* strand did not appear in the 1985 *Framework*, although some of its ideas appeared under statistics and probability. Discrete mathematics includes a cluster of related ideas, such as principles for counting arrangements of discrete objects (permutations, combinations, selections); other counting principles (the inclusion/exclusion principle, the pigeonhole principle); some basic and useful ideas from set theory (unions and intersections); the study of discrete structures (networks, graphs, and tree structures); recurrence relations (such as the Fibonacci relation, $F_n = F_{n-1} + F_{n-2}$); and the analysis of algorithms.

Discrete in this context means focusing on discrete and separate entities rather than on measures of continuous quantities; it does not mean that everything not continuous is to be considered discrete mathematics. Arithmetic with integers, for example, is treated under *number*, not under *discrete mathematics*.

Measurement

Student activity for this strand centers on the physical activity of measurement. Students use real tools to measure real objects and events.

Measurement is used in all occupations and in everyday life to compare. Numbers are assigned to quantitative aspects of the world by being compared to a scale of standard or non-standard units, such as inches, paper clips, kilograms, heartbeats, paces, or degrees Celsius. The measurement strand, by focusing on obtaining numbers through direct interaction with the universe, makes a physical connection between numbers and the world through the action of the student.

Work in measurement begins with comparison: bigger–smaller, heavier–lighter, warmer–colder. Students then create nonstandard units to help with the comparison. Later, they learn about standard systems of units, including the international metric system, especially units of time, distance, angle, weight, and temperature. Students also learn to combine units to find measures of other properties, such as area, volume, speed, acceleration, density, and pressure, and learn to apply other mathematics (e.g., trigonometry) to indirect measurement tasks.

During their careers students learn to choose suitable units, estimate, and allow for measurement error. They learn to judge the degree of precision appropriate for a given situation and the importance of accurate measurement and calculation; devise ways to assign numerical values to quantities they are interested in (a chocolate quality index or *Consumer Reports*-style ratings, for example); and even assess whether that process is appropriate.

Although this strand is closely allied with geometry through linear and angular measurement (angle, length, area, and volume), measurement involves more than using a ruler and a protractor. Measuring diverse quantities makes connections within mathematics, especially to statistics, and outside, to the natural and social sciences.

Number

Where do numbers come from and how do people use them? How does the system of numbers and operations work? The number strand weaves these two questions through the kindergarten through grade twelve curriculum.

Students learn about numbers from experience: they count how many, measure how much, and label objects in a collection. Over time they develop number sense, a sense for quantity in increasingly complex situations. Early on, students experience the power of mathematics to go beyond direct measurement and counting to answer *how many?* and *how much?* questions by using the basic operations of addition, subtraction, multiplication, and division.

The number system has power that is deeper than counting and collecting. Through exploration, usage, and reflective thought, students learn to make the system work for them. They learn how to use different kinds of numbers (integer, rational, real, complex, and vector) and what the basic operations mean. They learn about special numbers and properties (properties of 1, 0, π, reciprocals, and $\sqrt{2}$, for example). They learn how numbers relate to each other (order, inequality, betweenness, and density; factors, multiples, sums, and ratios). They learn to think with and communicate in the language of numbers (using various notations, verbal expressions with numbers, and number sentences). They become facile with techniques for computing (in their mind only or with the aid of pencil and paper or a calculator; through estimation or the use of algorithms). In the end students have an understanding of the system itself—how its simple elements give rise to a structure capable of representing relations among quantities in the real world and in the imagination.

This strand permeates all of mathematics. Numbers appear with all of the other strands—on coordinate axes and in labels on graphs and as the products of measurement, together with associated errors. Numbers describe scaling and proportionality and are the coordinates of data points and the representations of probabilities.

8. Logic and Language

The *logic* strand has been renamed *logic and language* to emphasize the importance of (1) language, in clarifying mathematical thinking and making valid arguments; and (2) mathematics, in giving natural language powerful tools for communicating complicated ideas clearly and precisely. In this way language connects mathematics to all other disciplines at all grade levels.

The *logic and language* strand focuses on the power of careful reasoning carried out in natural language to show things that are important but not obvious. However, formal deductive systems should not be introduced prematurely. (Very young students, especially when working together, can solve quite complex *informal* deductive problems.) Nor should explicit proofs of results that are intuitively obvious be emphasized. Proofs should make difficult things clear, not make clear things difficult. Nor should geometry be allowed to dominate what is done here because the ability to reason well and understand carefully worded explanations is too important to be associated with a single strand.

Ideas That Do Not Fit Within Strands

Analyzing mathematical content solely in terms of strands has two inherent limitations. First, that practice obscures the relative importance of topics within strands. Second, it obscures ideas that do not fit comfortably within any one strand because they cut across several. *Patterns*, mentioned previously, is one such principle. There are a number of other themes of great importance in school mathematics that fall into this category. In fact, they provide the rationale for introducing the other perspective on mathematical content, *unifying ideas*.

Unifying Ideas

A unifying idea is a major mathematical theme relevant in several different strands. Unifying ideas emerge when a higher-level view of content is taken. They tie together individual subjects, revealing general principles at work in several different strands and showing how they are related. Unifying ideas also set priorities. The curriculum is designed so that students develop depth of understanding in each of the unifying ideas at their grade level. Because deep understanding takes years of work, students will approach these large ideas repeatedly in many different contexts.

The concept of *proportional relationships* is an example of a unifying idea. Proportional relationships play a key role in a wide variety of important subjects, such as ratios, proportions, "per" and rates, percent, scale, similar geometric figures, slope, linear functions, parts of a whole, probability and odds, frequency distributions and statistics, motion at constant speed, simple interest, and comparison. Seeing the common principle operating in all these subjects is an important part of mathematical understanding; the use of *unifying ideas* allows us to focus on major underlying principles such as this one. Other examples of unifying ideas are *patterns* (mentioned previously), *equivalence*, and *algorithms*.

This *Framework* specifies just ten unifying ideas, only a few at each of the grade-level groupings: kindergarten through grade five, grades six through eight, and grades nine through twelve. (They will be described later under the content for each grade-level range.) Unifying ideas from earlier grade levels are still relevant at later stages; they are assumed, utilized, and extended and continue to form an active element of the mathematical work. But at each grade-level grouping, the most emphasis and prominence are given to the unifying ideas being newly introduced. Curriculum designers may cautiously add additional unifying ideas to those stipulated in this *Framework*. However, introducing too many unifying ideas would be self-defeating because the curriculum would tend to be fragmented instead of being unified. Only a small number of unifying ideas are focused on so that the essence of each is not lost and priorities for in-depth study can be defined.

Organization of the Curriculum

Although strands and unifying ideas describe the body of mathematics that California schoolchildren will learn, neither is an appropriate organizer for instruction. Strands bring breadth and balance to the curriculum; unifying ideas bring depth and connectedness. However, strands are incomplete if they appear only one at a time, and unifying ideas are too general and abstract to be the central focus of study.

An analogy related to cooking may help. Maria is a good cook. She integrates knowledge of ingredients and procedures and draws on her deep understanding of what tastes good and what makes one feel healthy. Her kitchen is organized by categories; the onions are near the potatoes, not with the spatulas. And there are procedures she knows from sautéeing to chopping to making gravy.

How can we teach cooking? It is tempting to sequence instruction according to the categories experts use. A course might be designed like a cookbook; that is, around types of foods and dishes and effective procedures. But if our students master legumes and then put them aside to study sauce-making, they will never get the big picture. Alternatively, if only the big picture—with categories like taste, nutrition, and presentation—is focused on, the students will be lost at the outset. A person unable to boil an egg cannot apply these complex concepts to his or her cooking. Instead, the curriculum would be more effective if every skill, each atom of knowledge, and each glimpse of the larger landscape appeared in the service of creating something useful. A person learns to cook by preparing meals. Culinary power cannot be developed in the absence of something good to eat.

To develop mathematical power, students need assignments in which the mathematics is more like a meal. The assignments must call for the use and development of diverse techniques and insights and be part of an intelligible (and tasty) whole. Students must do work that demonstrates all of the dimensions of mathematical power, interweaves strands, and relates to their own needs. Not just any collection of meals or assignments will do. They must make sense to the student on the one hand and support curricular goals on the other. Instruction needs to be organized in units of instruction—units whose mathematical subject matter *cannot* be represented simply in terms of strands and unifying ideas.

Units of Instruction

A typical unit consists of investigations, problems, and other learning activities, integrated with assessment, that develop depth of understanding and lead to complete work. Although instruction will take place outside these units, they are the building blocks of the program. Units should be long to accommodate substantial work by students; that is, the units will require one to six weeks for completion depending in part on the age of the students. Although a particular unit may build on earlier ones, each unit should have a fresh start and a distinct and unique character.

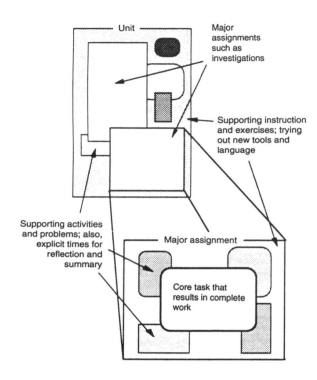

Unit

Major assignments such as investigations

Supporting instruction and exercises; trying out new tools and language

Supporting activities and problems; also, explicit times for reflection and summary

Major assignment

Core task that results in complete work

One way of envisioning the structure of a unit and its relationship to investigations, problems, and other activities

Coherent Subject Matter for Units

An instructional unit must be more than a collection of lessons and activities. The mathematical work must be *coherent*. The mathematics must be whole and complete and make sense in and of itself, and it must have a clearly formulated and understandable purpose. The work must include a wide variety of tasks, all clearly related to the primary goal, and must explore and eventually consolidate a set of related ideas that will be useful later. It must also develop an engaging context that invites students to participate, connects to their prior experience, and helps them see how mathematics is integrated in their lives.[1]

---

[1] It is easy to go too far in one direction. The "new math" of the 1960s, although rich in mathematical coherence, often failed to connect to students' (or teachers') prior experience or to convince them that it was interesting and relevant. On the other hand, "teddy-bear" units in which mathematics is sacrificed to being engaging or salable should not be created.

Each coherent piece interweaves strands, ensuring breadth, and each deepens one or more unifying ideas. In each, students will do mathematics and be responsible for the thinking. They will investigate rich and complex problem situations, conjecture, explore ideas, make connections among mathematical ideas, and generalize their findings. Such a body of curriculum is called a *unit*.

This turns out to be a pretty weird unit because all of the activities have to do with each other.

From an eighth-grader's reflection on the unit containing the proportional houses investigation described on page 18

If an idea for a unit arises, how does one know whether the idea is a good one? First, the mathematics and its relationship to the unit as a whole must be described. For example, a unit about *surveying* might identify similar triangles as part of the mathematics and then describe what they have to do with surveying. Given this description, the following six questions can be used as a guide to the suitability of the unit:

1. *Does the subject matter represent an* important *area of knowledge?* In a good unit the subject matter is important and relevant for all students in their mathematical education, other academic work, or everyday lives. It is not a subject justified only by criteria such as historical presence in the curriculum or internal consistency. Especially critical for most units is the development of a deeper understanding of one or more unifying ideas.

2. *Does the mathematics arise* naturally *in the unit?* In a good unit the mathematical ideas arise in a natural way out of the general setting and do not have to be forced. The unit shows clearly the power and purpose of the mathematical ideas around which it is organized.

3. *Are mathematical ideas* integral *to an understanding of the unit?* In a good unit the mathematical ideas are essential for a clear understanding of the general subject matter of the unit. They are not peripheral.

4. *Does the unit have natural* coherence? In a good unit the subject matter is a unified exploration of a single theme. Whether the theme is fully mathematical (as in more abstract units) or is based on a subject outside mathematics (as in more concrete units), the single focus gives the unit a natural coherence. A good unit is more than a collection of related topics.

5. *Does the mathematics in the unit have important* internal connections *within mathematics?* In a good unit students see the mathematical ideas and exploit connections to other areas of mathematics. The ideas do not appear only in this unit and nowhere else. The best units draw on ideas from different strands and develop more than one unifying idea.

6. *Does the unit provide* external connections *to work in other disciplines?* In a good unit the mathematics is relevant and useful for work in other disciplines, such as history, science, vocational courses, or economics, or builds on work students are already doing in these disciplines or both.[2]

Subject matter that is more concrete. Some units can be devoted to subjects outside mathematics that are important for a quantitative understanding of the world and can benefit from an analysis in terms of mathematics. In these units *students use mathematics to make sense of something important that is not mathematics.* For example, students might study maps and scale drawings.

In a sense concrete units represent mathematical *applications.* In traditional mathematics textbooks, however, applications appear *last,* after the mathematics; their purpose is primarily to illustrate. In contrast, students can often make more sense of mathematics when the application appears *first,* introducing and motivating the mathematics. The goal is *both* to learn about the subject matter of the application *and* to understand the mathematics used in its description.

Mathematical applications are too often presented as unessential afterthoughts that receive only fragmentary treatment. Students are given the impression that the only legitimate purpose of mathematics courses is to teach *real* or *pure* mathematics. This view must be challenged because students need

---

[2]See the section on integrating mathematics with other subject areas, pp. 101–2.

the useful and coherent understanding of basic applications of mathematics.

Units based on concrete topics introduce mathematical ideas through a subject that is understandable at an intuitive and informal level when approached nonmathematically. When mathematical concepts appear in the unit, they help to describe the subject and have a clear purpose. Students are thereby motivated to understand the mathematical concepts and a readily accessible, concrete illustration of the concepts. The result is that students learn and retain more mathematics and come to understand more effective uses of mathematics.[3]

Some may object that more concrete or applied topics such as *growth* or *motion* belong in science courses or other courses and not in mathematics courses because mathematics is not focused on initially. However, the rigid separation of mathematics and science benefits neither subject. Concrete units inspire useful work that typically is not done in either mathematics or science. They are not so much specific applications in science as they are prototype applications needed for a general quantitative understanding of the world.

Subject matter that is more abstract. Students will also study subjects organized around abstract mathematics. They work toward an understanding of a coherent, unified, and useful aggregate of mathematical ideas and learn their way around mathematics as a powerful, self-contained intellectual discipline.

In a sense abstract units represent *pure* mathematics but without an emphasis on theory and proof. Instead, pure mathematics here simply means mathematics apart from its applications but does not preclude the use of illustrative examples, concrete models, or manipulatives, which are appropriate in all kinds of units. Here they further the primary goal of learning something that is entirely mathematical.

Abstract units fill two roles. First, they develop purely mathematical ideas for their own sake. Second, they provide a summarizing and integrating function, letting students reflect on and connect related topics from previous units. This work can also provide students with mathematical understandings,

---

[3]Use of concrete materials or manipulatives is valuable for instruction in all kinds of units, whether more concrete or more abstract. However, the use of such a learning tool is different from the study of a concrete subject such as *growth* or *motion*. In a concrete unit the goal is not to use concrete materials as a learning tool but to learn something about specific subject matter that is concrete or applied.

tools, and techniques they will need in later units. For example, a unit on *the meaning of multiplication* is an example of an abstract unit in elementary mathematics. In such a unit students would encounter situations that call for multiplication. They would calculate in their mind only or with the aid of paper and pencil, manipulatives, or calculators; but, more important, they would come to understand the variety of ways in which multiplication is useful. The point of the unit is understanding, and an inevitable by-product is practice in computation. Work in such a unit would prepare students for future work; for example, about area and perimeter or about buying, selling, and making change.

At the high school level, for example, once students have had some experiences with concrete contexts that naturally produce functions, they are ready for a visit to the abstract perspective, *families of functions*. There they will see that many of the functions they have studied are the same and that transformations can apply to all functions. In such a unit they may study functions of the form $f(t) = Cb^t$, which they can use to study population growth or geometric sequences.[4]

The idea is not simply that exponential functions have been covered because students have done some exercises and problems involving manipulation of exponential expressions. Such a treatment would not develop the kind of significant understanding of exponential functions needed to see their relevance to population growth or geometric sequences. The work done in a unit based on the *families of functions* is broader than a coverage of mechanics; it gives students a conceptual overview of major principles mixed appropriately with computational facility in specific problems.

Unit Structure

There is no single correct way to structure units or decide what they should be about. Designers should create units that develop all of the dimensions of mathematical power, support students working collaboratively and independently, develop students' positive dispositions towards mathematics, and take into account historical, societal, and career information. In addition, assessment should be integrated with instruction in the unit. What is implied here is that the units are built around

---

[4]Presumably, these high school students have had concrete experience with patterns of growth to prepare them for this work; for example, their study of *growth* in middle school and prior abstract experience with linear functions.

large pieces of work that can be assessed for completeness; for example, projects such as investigations.

Just as units bring several interrelated mathematical ideas together, the major assignments and activities within the unit—the problems and investigations—are rarely based on a single concept or objective. Most require several days to complete and (when students are old enough) take advantage of the time available to students *outside* the class. Students can spend more time as needed. Many assignments are open; that is, students are asked to generate questions, explore a variety of approaches, and share and interpret findings.

Another question arises in the design of units. How big should they be? Consider a unit on *indirect measurement*. What should that unit include? Finding the distance to the top of the flagpole? The distance to the moon? The mathematics of similar triangles? Trigonometry? What about inferring a growth curve for students—how tall they are at each grade—by looking at the distribution of heights among students in different grades? Although that task requires very different mathematics, it involves indirect measurement. Should it be in the same unit? Or should the subject matter of the unit instead be *surveying*? It depends. Units can be broad and suitable for revisitation many times or narrower and specific—but no less suitable. The purpose of a unit determines its scope.

Strands and Unifying Ideas

Instructional units must be carefully designed to be unified wholes dealing with important subject matter. Strands and unifying ideas represent perspectives on mathematical content. But how do they fit into units? Consider, for example, units based on *growth*, a suitable topic for grades six through eight.

Strands come into the picture as a way of thinking about the *breadth* of the kinds of mathematical ideas that appear in particular units. An effective unit on *growth* needs to use concepts from different strands. *Number, functions, algebra, measurement,* and *statistics* are all relevant to the mathematics of growth, and any unit needs to involve more than one of these strands in a major way.

Strands also help us achieve *balance* in the units used through the year. It would be artificial and unproductive to try to include ideas from every strand in every unit. Yet in the course of a year's time, most if not all strands need to make a significant appearance.

Unifying ideas help to achieve *depth* and *connectedness* in the curriculum by focusing on mathematics that reappears in different units. For example, when situations involving constant rate growth are explored in a unit, many opportunities exist for discussing the unifying idea of *proportional relationships;* that is, the amount that something grows is proportional to the time it has been growing. Because students can portray growth vividly by using models, geometric diagrams, tables, graphs, and functions, *growth* units can also illustrate the unifying idea of *multiple representations.* Finally, many examples of growth allow students to see patterns and make mathematical generalizations based on those patterns. In this way examples of growth add to the understanding of the unifying idea of *patterns and generalization.*

Not every unifying idea has to be brought equally into every unit. Rather, opportunities that arise naturally in a particular unit can be used to emphasize one or two unifying ideas. At the same time one measure of a unit's appropriateness is the degree to which it addresses the unifying mathematical ideas for its grade level.

Because strands and unifying ideas represent *perspectives* on curriculum and not a *division* of curricular content, a single word or phrase may represent more than one of these perspectives when interpreted in different ways. For example, *functions* is one of the eight *strand*s of mathematical content. Yet looked at more broadly, the theme of *functional relationships* also operates as a *unifying idea* for grades nine through twelve, spanning several strands. Thus, *algebra* provides expressions for describing many basic functional relationships; transformations in *geometry* are a kind of functional relationship; *discrete mathematics* has recurrence relationships; *statistics and probability* uses distributions, which share many of the properties of functions; and so on.

Of course, other conceptual schemes and categorizations of mathematics content are possible. Seven or nine strands could have been specified—or four ways of characterizing content instead of two. The important thing is that the scheme and categories provide ways of constructing the goals of the program so that they can be used to make choices about curricular content: breadth and balance, priorities for depth, and meaningful coherence.

A Year's Work

A year's work will consist primarily of a set of units that not only are internally coherent but also develop progressively the depth of understanding of unifying ideas, the understanding and use of mathematics laid out in the NCTM *Standards*, and all the dimensions of mathematical power. An overall description of a year's work should outline the length and sequence of units. However, flexibility being important, student and teacher choices should be possible because they will make the mathematics more meaningful.

Trying to do everything important in every year would, however, result in superficial treatments of little value. Organizing instructional units around coherent, meaningful subject matter gives priority to depth and connections, even though some traditional content may not be covered. Choices must be made about what material to treat each year—choices that will ensure that what is addressed is what is believed to be the most important mathematics for each grade level. In the descriptions of mathematical content by grade level, the NCTM standards help balance curricular choices about depth and connections with the scope of mathematical content.

Possible ideas for units (on pages 109–16, 125–28, and 169–74) are too few to specify the whole curriculum. Program authors, school districts, and teachers may use those ideas if they wish but must create additional units to complete the curriculum content. A flexible ensemble of units makes up a varied and balanced diet of mathematical work. The units provide balance across the strands and ensure substantial attention to the unifying ideas.

Some units may be distributed over time. Students are thereby enabled to have experiences with activities or ideas from the unit before or after working on the main part of the unit. For example,

97

in a unit on game strategies, students will play some of the games many times, perhaps over several months, before analyzing them. In another unit students may be asked to collect data over an extended period. At times it may make sense for students to reflect on their investigations at a later time rather than concentrate all the experiences at once.

Students need a variety of learning experiences. Programs can supplement long, in-depth units with short units that introduce or summarize mathematical tools, techniques, or ideas in a straightforward way. Occasional lessons not directly connected to the subject being studied can develop skills, provide variety, and activate connections to other learning. Curriculum designers will and should use their creativity and experience to supplement the structure of this *Framework* as they put its ideas into instructional materials and activities. Nonetheless, the basic building blocks of the curriculum should be the content described in this document.

Material Between Units

Not everything in the curriculum should be tightly organized in units. Smaller fragments of time can be reserved for activities unrelated to units or unifying ideas. In these smaller fragments students can be given variety: favorites from years past, problems of the week, short exercises, and so on. Instructional materials should contain some of these; individual teachers can provide others.

In addition to these mathematical fragments, students should regularly be lifted above the chain of units to reflect on what they have done and summarize what they have learned. Summarizing and reflecting will be a normal part of every unit, of course, but should also be a major activity—looking back on several units— several times during the year. Students of all ages can reflect on their learning, writing and talking about it. Even older students can be expected to begin to connect their mathematical experiences on their own and to synthesize deeper understanding—but only if they have time set aside to do so. This reflecting and summarizing might include collaborative discussion topics or reflective writing assignments. Students also benefit from summary presentations (by teachers or peers) of mathematical techniques, notation, rules, and so on (organized for easy reference rather than for instruction) and of the connections among various units. Students and teachers can also spend this time looking over portfolios and reviewing assessment schemes.

Selecting Units That Develop Mathematical Power

Programs should be organized for coherence and cumulative effect. How should a year's worth of the program be constructed? As pointed out in a previous section (pages 95–96), it is important to ensure that the year's content is sufficiently broad by seeing that each strand is adequately represented and sufficiently deep and directed by evaluating the contributions units make to understanding the unifying ideas. Consideration of the other dimensions of mathematical power are equally important in deciding what to include. With regard to mathematical thinking, for example, it is important to make sure that students generate (conjecture, construct, design, formulate, predict) and critique (analyze, evaluate, verify) their own ideas and the ideas of others. These activities should occur in a wide variety of contexts and around a broad selection of deep mathematical ideas.

Over the course of each year, students should employ a full range of tools and techniques, including concrete materials and physical models, mathematical symbols, diagrams and drawings, calculators and computers, measuring instruments, algorithms, and mathematical notation. Over several years students should have so much experience with different tools and techniques that they can make intelligent selections among them. This result will occur only if all tools and techniques are familiar, frequently useful in student work, and continually available.

Finally, opportunities for communication need to be varied in audience and purpose. Thus, over the course of a year, each student is expected to present and explain his or her thinking to an expert, a layperson, a parent, a teacher, a peer, and a younger sibling to inform, explain, teach, and persuade. Students should communicate orally and in writing; use pictures, graphs, tables, and symbols; and produce projects, journals, presentations, and performances.

Across the Years

The design of a mathematics program transcends grade levels. Just as a unit is more than a collection of mathematical lessons and a year's work is more than a collection of units, a

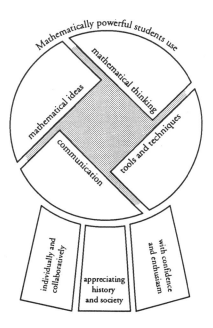

"We must know that the understandings we seek to help the student gain are developed, elaborated, deepened, and made more complete over time. We must provide a variety of opportunities to explore and confront any mathematical idea many times."

Mathematics Model Curriculum Guide (Guiding Principle 3), p. 13

total program (kindergarten through grade six, kindergarten through grade eight, grades six through twelve, or grades nine through twelve) is more than a collection of grade-level programs. The accompanying diagram shows the overall path a student will take in mathematics from kindergarten through grade twelve. (For a detailed discussion of the high school core sequence, see Chapter 5.)

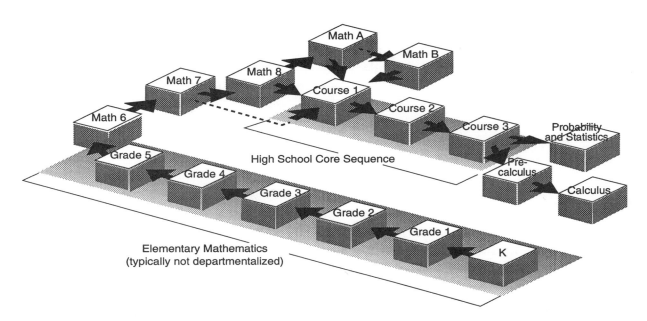

A student's mathematical inquiries and understandings should deepen each year. For example, *the mathematics of growth* is studied in grades six through eight. During these three years a unit (and perhaps two or three) will probably be devoted to an understanding of some aspect of growth and the mathematics needed to describe it. It makes sense for students to revisit the most important mathematical ideas year after year. Students may even repeat the same activity at different grade levels. But if an activity is revisited, it may be more complex; or the students may probe the mathematics differently or more deeply.

In the design of a mathematics program, two related challenges are faced—revamping the curriculum to focus on more important mathematics in all of its dimensions and developing every student's mathematical power. A new, multidimensional curriculum will be accessible to students who are mathematically less experienced and, at the same time, will be interesting to those who are more sophisticated.

Looking to the Future: Integrating Mathematics with Other Subjects

Like writing, mathematics has always had a double life; it is a discipline in its own right and also plays a central role in many other disciplines. The role of mathematics in science, history–social science, and vocational education is well known. Mathematics also enriches the visual and performing arts and even physical education. And, as an extension of natural language, it provides a context for learning languages.

As mathematics pervades all facets of society, its integration with other school subjects becomes an important goal. Over the next ten years, curriculum designers at all levels, including teachers, can make important progress toward achieving that goal. The structuring of mathematical content in meaningful units represents a large step toward making integration practical. Many of the suggested units described in this *Framework* have interdisciplinary focus.

Coherent units will be especially useful in creating mathematics programs for newly designed vocational programs.[5] The formulation of vocationally based units ensures that, even with crowded schedules, students in vocational programs will have solid options in mathematics. For example, a unit developed around *decisions in health care* might include the study of risk management, probabilities in treatment choices, resource allocation problems, and inference in medical research. Such a unit would benefit future physicians, nurses, pharmacists, medical technicians, and other employees of the health care industry—as well as every student who becomes a consumer of health care. Other examples of vocationally based units might include *machines, electronic technologies, business projections* (including use of spreadsheet software), *exploration of data bases*, and *construction*.

Yet integration of mathematics with other subjects is not easy. Haphazard integration can result in serious watering down of the curriculum. The following guidelines address

---

[5]Many schools are restructuring vocational programs to improve offerings for students who have been enrolled in general education courses.

important issues in designing an integrated assignment, unit, or course:[6]

1. *The assignment, unit, or course should advance learning in each of the subjects integrated.* For example, a sixth-grade science unit that uses third-grade mathematics hardly counts as instruction in mathematics or integrated curriculum. A truly integrated unit would advance both subject areas by using sixth-grade science *and* sixth-grade mathematics.

2. *The assignment, unit, or course should focus on curricular goals that are central to all of the subjects integrated.* When a unit is being designed that integrates business and mathematics, for example, both frameworks should be consulted. The unit should develop unifying ideas from mathematics as well as key themes from business. It is a costly mistake to devote time to ideas that are only peripherally important to the subjects being integrated.

3. *An integrated assignment should be assessed from more than one perspective.* For example, in a unit that integrates mathematics and economics, students might be asked to analyze the issue, Is inflation fair? The economics teacher would assess the students' understanding of how inflation works, its economic consequences, and its interaction with income levels. The mathematics teacher would assess the students' analysis and representation of relationships among the variables of cost, income, and investment earnings as a function of income. And the English teacher might assess the writing quality of the analytic essay.

It is not too soon to venture toward more integrated instruction. At the elementary level the demands on learning time already require integration. At all levels many ideas and skills can be developed in an integrated context. A good place to begin is with a few integrated assignments. As design issues are understood better, units and then entire courses can be developed. In particular, efforts should be made toward producing alternative versions of the core sequence for grades nine through twelve, including interdisciplinary themes, taking care that the alternatives do not water down the standards for mathematical power.

---

[6]The design guidelines discussed on pages 89–96 of this *Framework* should also be consulted when formulating interdisciplinary units.

Mathematical Content in Kindergarten Through Grade Eight

C h a p t e r **4**

This chapter contains a description of the mathematical content for kindergarten through grade eight. One section is devoted to the elementary grades and another to the middle grades. Within these sections the appropriate NCTM standards are listed, and the unifying ideas for that grade-level range are described. (A description of the strands appears in Chapter 3, pages 80–87.) Finally, subject matter for potential units of instruction is suggested.

Content of the Elementary Program

In the following discussion of mathematical content for the elementary grades, it is assumed that the reader is familiar with the general description of content in Chapter 3 beginning on page 77.

Strands

Functions

Algebra

Geometry

Statistics and Probability

Discrete Mathematics

Measurement

Number

Logic and Language

Unifying Ideas

How many? How much?

Finding, Making, and Describing Patterns

Representing Quantities and Shapes

Suggested Units for the Elementary Grades.

Attributes and Classification

Understanding Number and Numeration

Understanding Arithmetic Operations

Dealing with Data

The Process of Measurement

Measuring Geometric Figures

Locating and Mapping

Visualizing and Representing Shapes

Exchange

Games and Rules

Sharing

NCTM Standards for the Elementary Grades

The NCTM standards for kindergarten through grade four are excerpted here. By the end of the fourth grade, students should be able to do what is described in these standards. Those concerned with grade five may wish to consult the Middle School Program, where the standards for grades five through eight appear (pages 118–21). These excerpts are not intended to be self-explanatory; therefore, the reader is advised to see the full discussion in the NCTM *Standards*. (See also Appendix E in this *Framework*.)

Standard 1: Mathematics as Problem Solving

In kindergarten through grade four, the study of mathematics should emphasize problem solving so that students can:

- Use problem-solving approaches to investigate and understand mathematical content.
- Formulate problems from everyday and mathematical situations.
- Develop and apply strategies to solve a wide variety of problems.
- Verify and interpret results with respect to the original problem.
- Acquire confidence in using mathematics meaningfully.

Standard 2: Mathematics as Communication

In kindergarten through grade four, the study of mathematics should include numerous opportunities for communication so that students can:

- Relate physical materials, pictures, and diagrams to mathematical ideas.
- Reflect on and clarify their thinking about mathematical ideas and situations.
- Relate their everyday language to mathematical language and symbols.
- Realize that representing, discussing, reading, writing, and listening to mathematics are a vital part of learning and using mathematics.

Standard 3: Mathematics as Reasoning

In kindergarten through grade four, the study of mathematics should emphasize reasoning so that students can:

- Draw logical conclusions about mathematics.
- Use models, known facts, properties, and relationships to explain their thinking.
- Justify their answers and solution processes.
- Use patterns and relationships to analyze mathematical situations.
- Believe that mathematics makes sense.

Standard 4: Mathematical Connections

In kindergarten through grade four, the study of mathematics should include opportunities to make connections so that students can:

- Link conceptual and procedural knowledge.
- Relate various representations of concepts or procedures to one another.
- Recognize relationships among different topics in mathematics.

105

- Use mathematics in other curricular areas.
- Use mathematics in their daily lives.

Standard 5: Estimation

In kindergarten through grade four, the curriculum should include estimation so students can:

- Explore estimation strategies.
- Recognize when an estimate is appropriate.
- Determine the reasonableness of results.
- Apply estimation in working with quantities, measurement, computation, and problem solving.

Standard 6: Number Sense and Numeration

In kindergarten through grade four, the mathematics curriculum should include whole number concepts and skills so that students can:

- Construct number meanings through real-world experiences and the use of physical materials.
- Understand the numeration system by relating counting, grouping, and place-value concepts.
- Develop number sense.
- Interpret the multiple uses of numbers encountered in the real world.

Standard 7: Concepts of Whole Number Operations

In kindergarten through grade four, the mathematics curriculum should include concepts of addition, subtraction, multiplication, and division of whole numbers so that students can:

- Develop meaning for the operations by modeling and discussing a rich variety of problem situations.
- Relate the mathematical language and symbolism of operations to problems and informal language.
- Recognize that a wide variety of problem structures can be represented by a single operation.
- Develop operation sense.

Standard 8: Whole Number Computation

In kindergarten through grade four, the mathematics curriculum should develop whole number computation so that students can:

- Model, explain, and develop reasonable proficiency with basic facts and algorithms.
- Use a variety of mental computation and estimation techniques.
- Use calculators in appropriate computational situations.
- Select and use computation techniques appropriate to specific problems and determine whether the results are reasonable.

Standard 9: Geometry and Spatial Sense

In kindergarten through grade four, the mathematics curriculum should include two-dimensional and three-dimensional geometry so that students can:

- Describe, model, draw, and classify shapes.
- Investigate and predict the results of combining, subdividing, and changing shapes.
- Develop spatial sense.
- Relate geometric ideas to number and measurement ideas.
- Recognize and appreciate geometry in their world.

Standard 10: Measurement

In kindergarten through grade four, the mathematics curriculum should include measurement so that students can:

- Understand the attributes of length, capacity, weight, area, volume, time, temperature, and angle.
- Develop the process of measuring and concepts related to units of measurement.
- Make and use estimates of measurement.
- Make and use measurements in problem and everyday situations.

Standard 11: Statistics and Probability

In kindergarten through grade four, the mathematics curriculum should include experiences with data analysis and probability so that students can:

- Collect, organize, and describe data.
- Construct, read, and interpret displays of data.
- Formulate and solve problems that involve collecting and analyzing data.
- Explore concepts of chance.

Standard 12: Fractions and Decimals

In kindergarten through grade four, the mathematics curriculum should include fractions and decimals so that students can:

- Develop concepts of fractions, mixed numbers, and decimals.
- Develop number sense for fractions and decimals.
- Use models to relate fractions to decimals and to find equivalent fractions.
- Use models to explore operations on fractions and decimals.
- Apply fractions and decimals to problem situations.

Standard 13: Patterns and Relationships

In kindergarten through grade four, the mathematics curriculum should include the study of patterns and relationships so that the student can:

- Recognize, describe, extend, and create a wide variety of patterns.
- Represent and describe mathematical relationships.
- Explore the use of variables and open sentences to express relationships.

Integrating Strands in the Elementary Grades

The eight strands of mathematics identified in this *Framework* are for kindergarten through grade twelve. For kindergarten through grade five and for other grade spans, the program must interweave and integrate the strands. Fortunately, many relatively simple mathematical activities draw on several strands. Because the strands are intended to be interwoven, students need not have experiences with each strand separately. In kindergarten through grade five, some strands, particularly *algebra*, *functions*, and *discrete mathematics*, seldom appear alone because it is not appropriate to deal with that material formally.

For example, the elementary activity that follows—a counting activity—draws on three strands: *number*, *functions*, and *algebra*:

A class of first graders practices counting by twos. Together, they count out loud the number of shoes worn by a group of students. While the class counts, one student records the counts on the board in a T table, and another student records the counts in the number sentence—2 shoes + 2 shoes + 2 shoes + ... = 16 shoes.

Unifying Ideas for the Elementary Grades

How many? How much? Quantification permeates the elementary mathematics curriculum. Children begin to count and measure things before they start school. They want to know how tall they are or how many cookies are on a plate. They compare to see who has more cookies or stand back to see who is taller.

A count tells *how many* things are in a specified group. A measure tells *how much* of a specified attribute something has. Generally, counts are discrete, and measures are continuous. Both counts and measures identify quantities; therefore, both are identified by unit labels, such as oranges or centimeters.

As children count and measure, they learn how the system of numbers is structured. They combine or compare counts and measures to determine other quantities, make visual estimates of physical amounts, and estimate mentally the results of calculations. They develop a sense of range from few to many and from small to very large and can express their sense numerically. They understand the effect of the operations on the size of quantities operated on.

Finding, making, and describing patterns. The search for patterns is fundamental to mathematics at all grade spans. Patterns help children to see order and make sense of the underlying structures of things, situations, and experiences. Patterns help children predict what will happen.

Children encounter many patterns in the world: in designs, in rhythms they hear, and in recurring events, such as the days of the week or the seasons. They will see mathematical patterns in both numbers and shapes. Some patterns are sequential (one thing follows another) and may be repeated periodically (such as the ones' digit when counting by twos or threes); or they may continue to grow (such as the powers of two). Other patterns are spatial, such as tessellations or symmetrical designs, or both spatial and numerical, such as the square and triangular numbers or the multiples of 3 circled on a 0–99 chart.

Students use mathematical patterns to make sense of situations involving functional relationships. In kindergarten through grade five, students look for patterns in all they do, creating and extending patterns and looking for the same or similar patterns in a variety of situations.

Representing quantities and shapes. Children forge a link between their concrete everyday experiences and their understanding of mathematical abstractions through many experiences in representing quantities and shapes. Representation helps children remember an experience and communicate it to others. Representing quantities and shapes is also an aid to learning; it can help children make mathematical sense of what they are doing.

Children's first representations of quantities and shapes are very simple and unsophisticated. They use physical materials to represent counts or to model other objects (such as a block model of a building or desk). They draw a picture and talk about what they have done. As children develop, however, their drawings become more complex, and they begin to describe their experiences in writing. They use numbers, number expressions, and sentences to record situations involving quantities. They see how organizing data into graphs and tables—and using diagrams and maps instead of words— makes mathematical ideas easier to comprehend.

Suggested Units for the Elementary Grades

The sections that follow contain suggestions related to subject matter for units of study in the elementary school. But what will students in different grades *do*? Students' mathematical knowledge and power change greatly between the time they enter kindergarten and the time they leave grade five. Appendix B describes sample experiences that students might have with the mathematics organized around the ideas presented here.

An empowering elementary school curriculum will have students revisit many of the issues presented, possibly every year. Although this *Framework* does not require specific units, see the section "Dealing with Data" (page 111) for an example of how units might be developed in the elementary grades.

Attributes and classification. Children's experiences with sorting and classification in kindergarten through grade five provide the foundation for more formal work later on with logical thinking. Developing the concept of an attribute is

critical because it helps answers such questions as the following: What are the characteristics of something? How are things alike, and how are they different? How can we group things? Which group and why? This subject, like others, has roots in common with important ideas in science. Assignments and units of study that integrate mathematics and science can make potent connections for students.

As they work with attributes and classification, children reason mathematically in many ways. They analyze and generalize the characteristics of attributes; use multiple attributes; define attributes to clarify ideas; and record and apply classification systems to decisions in problem-solving situations. Children communicate their ideas mathematically through language, representations, diagrams, pictures, and explanations.

Understanding number and numeration. When children begin school, they are already using numbers in their daily lives. They count small groups of things and know their own age. Many know their own phone number. Their experiences in school will gradually expand and deepen their understandings of number and numeration. They will develop number sense and facility in doing things with numbers. By the time they leave grade five, students will understand how to:

1. Use numbers to show quantities of whole objects and parts of wholes.
2. Represent very large and very small quantities through grouping and place value.
3. Represent a number in many ways, including equivalence.
4. Use numbers to define quantities, make comparisons, and identify locations.

Understanding arithmetic operations. Understanding the four basic arithmetic operations has always been essential to elementary mathematics programs. However, the focus of attention has shifted from proficiency with computational procedures to attempts to make sense of and use addition, subtraction, multiplication, and division appropriately. Overemphasis on recall and speedy execution of algorithms can interfere with students' understanding of mathematical concepts. Children often become confused in trying to memorize a step-by-step procedure unconnected to experience. For example, in working on a multidigit regrouping subtraction problem, children commonly subtract the smaller digit from the larger one.

Students need to be able to use the operations to find answers to problems. They need the basic facts, which are essential for approximating answers and determining whether answers are reasonable, and need to decide whether calculating in their mind only or with the aid of paper and pencil or a calculator is suitable for a particular task. Their understanding of the operations develops slowly as they grapple with ways to solve real mathematical problems encountered in different situations. In time students will begin to recognize generalizations and procedures that make sense. (See the section titled "Procedures for Manipulating Numbers and Symbols" in Chapter 2.)

Understanding the four basic operations involves knowing what actions are represented by the operations, how these actions relate to one another, and what operations are appropriate in particular situations. This kind of fluency with the four basic operations is also a prerequisite for exploring mathematics from all of the strands.

Dealing with data. Students collect, analyze, describe, organize, and interpret data so that they can make decisions and predictions on the basis of the information. Key concepts include the following:

1. Data can represent information from the real world.
2. The ways in which questions are asked can influence the information gathered.
3. Data can be organized, represented, and summarized in a variety of ways. (A sample question might be: What is the best way to show these data?) Some ways of displaying certain data are more useful or appropriate than others.
4. Sometimes a representative sample may be used instead of data from an entire group.

The chart on the following page illustrates how units could be included each year in a program. The focus of the units changes from year to year as the students mature and deepen their understanding of mathematics.

The process of measurement. Whatever students measure and whatever the attribute (such as length, weight, or time), they will select an appropriate unit, standard or nonstandard. They compare that unit with the object being measured and determine the number of units by counting, reading a measurement instrument, or using a formula. And most measures are approximations, not discrete counts.

In an attempt to show how coherent units develop thematically across the grades, six possible units on "Dealing with Data" are presented here:

K

What Do You Choose?—This unit is distributed over time to allow kindergarten students many opportunities to experience similar activities. Students are presented with various situations and are asked first to choose between two items and later among three or more items. For example, at snack time they are asked to select either an apple or an orange. They line up their choices to compare which item was selected more, count the number in each group, and draw conclusions about their selections.

Grade 1

Counts and Amounts—This unit combines data collection and display with students' number work. Students count different objects in the room, such as glue bottles, library books, windows, and chairs. They group their counts by tens and ones and then illustrate their findings in charts and graphs.

Grade 2

Comparing Collections—Students bring to school or describe their collections. They examine individual collections, describing, sorting, classifying, counting, and graphing the items in them. They graph the kinds of collections students in their class have.

Grade 3

Displaying Data—Students investigate different ways to display information they have collected via tables and graphs. They discuss options about displaying data; for example: How might data be grouped or categorized? What makes a graph or table easy to understand and interpret? How can the same data be displayed in different ways and what are the advantages and disadvantages of each way?

Grade 4

Looking at Us—The purpose of this unit is to provide students with their first sustained experience with data analysis—from choosing a question, collecting data, and analyzing them to representing data in a graph and interpreting them for their classmates. The students brainstorm a list of characteristics describing the children in the class, and pairs of students each select a characteristic they want to research. They share the results of their research in graphs and reports and in a school newsletter article.

Grade 5

Sampling—Students examine survey findings from the media and speculate how the findings were determined. They conduct a number of experiments (with cubes and beans in a bag) and compare results from a number of samples with the entire "population" of cubes and beans. They discuss why sampling is necessary. They decide on a survey question of interest to the school. They prepare a random sample and then compare the results with a written survey of the entire school. They discuss differences between the two surveys.

Beyond the building of measurement skills, it is critical that the contexts for measuring appear sensible to the students, who must see a purpose or use for any measuring they do. In addition, curriculum designers should attend to the mathematics behind measuring and to the mathematics students need in the measurement process. When is it appropriate to add angles? What is meant by subtracting measurements of time? What is meant by multiplying two lengths? Are two weights *ever* multiplied? What reasoning must be used to measure the thickness of a wall? What about the thickness of a piece of paper?

In addition, students make estimates and perform measurements to refine them. For example, guessing how many children would fit into a classroom is just a guess that becomes better informed through measurement, reasoning, and calculation.

Measuring geometric figures. Often, things such as the following need to be known: Will an object fit into the space available? How much will the space hold? How much will be needed to cover or surround the object? Is the most efficient use being made of space or material?

Students first approach the concepts of volume, surface area, and perimeter empirically; they measure by filling, covering, and stretching string rather than by calculating. Elementary school students can deal with optimization problems (what is the most rice we can fit into a box made from one piece of paper?) if they are allowed to work in groups and use concrete materials.

As students measure more and more geometric figures and other objects, they uncover patterns and see relationships. This work helps to lead eventually to generalizations such as formulas for area and perimeter.

Locating and mapping. Locating and mapping involve questions such as the following: How do I tell someone where something is? How do I get from here to there? How far is it? How can I represent the physical world? How do I interpret representations of the physical world?

Locating things and using maps, common applications of mathematics encountered in everyday life, connect to school subjects such as geography and science. The study of these ideas begins with concepts as fundamental as near and far and left and right. It proceeds to measurements of length, area, and distance; to concepts of north and south, map scales, spatial visualization, models, and coordinate grids; and to the giving and receiving of instructions.

"Premature use of instruments or formulas leaves children without the understanding necessary for solving measurement problems."

—*Curriculum and Evaluation Standards*, p. 51

113

When students describe and create models of their house or their classroom, they deal with buildings and their shapes. Students explore indoor spaces and begin thinking about the use of space and the planning of an environment. When they use verbal descriptions, maps, and floor plans of places they can walk to (to give directions to a treasure, for example), they come to understand the correspondence between the representation and the reality. As they go further afield, using maps of the town, of California, and of the United States and world maps and globes, they develop their ability to abstract their concrete understanding and use maps meaningfully.

Visualizing and representing shapes. How can I represent a three-dimensional object on paper? How can I build a model by looking at a picture of the object? How does your view of an object differ from mine because we are in different locations? What shapes can I see in different objects? How do I make a symmetrical design? Which shapes will tile a floor (or tessellate)? These questions involve spatial sense.

Through direct experiences with the three-dimensional world and then the two-dimensional world, students become aware of the properties and relationships of geometric figures and are able to analyze shapes. They notice symmetry in the designs they create and in the patterns they see around them. In addition, they investigate how different shapes and patterns look when they are rotated or reflected in a mirror and draw, build, or describe various shapes from a variety of perspectives.

Exchange. Exchange is a phenomenon that can occur when two objects have the same *function* or the same *value* from someone's point of view. In particular, people buy things, trade, substitute equivalent measures, and convert measurements. They explore what they can substitute or exchange for something else. The study of exchange is rich in mathematical and contextual connections. Do two red trapezoids "make" a yellow hexagon? Exchange deals with the number and spatial sense that enable the student to understand these forms of equivalence.

Students deal with money frequently—what coins and bills to use and what change to receive. Money questions are common for us and for children: What should I buy? Where will I get the money? How long do I have to save to buy it? What do I save if it is on sale? Is it worth the price?

Activities related to exchange include experiences that allow students to show their understanding through language, diagrams, pictures, and the manipulation of real objects: coins, blocks, and measuring tools. In the context of exchange, students also add and subtract and use their understanding of place value, fractions, and decimals. In addition, they develop fundamental geometrical knowledge, spatial visualization skills, familiarity with standard units, and the flexibility to use nonstandard ones. Experience with exchange builds a good foundation for understanding equations and operations on equations.

Games and rules. Students play games. They participate as individuals and as team members; indoors and outdoors; with equipment and without. The games themselves require skill or luck or both, and they may be competitive or not. For all their variety all of the games have rules, both implicit and explicit. The rules connect games to the mathematics of language and logic.

Students learn rules for games and explain them to others, orally and in writing. They create rules for new games of their own and modify rules for existing ones. They invent, test, and compare strategies for games. They analyze games. Which games require the most skill? Which are most dependent on luck? They even discuss what makes a game enjoyable and what makes something a game.

Games that involve luck also give students experience with probability. In the elementary school years, students are not required to compute complicated theoretical probabilities. Instead, they have a rich variety of experiences in which they sometimes observe chance informally and sometimes record extensive data and report empirical results. For example, they make spinners with different-sized sectors and notice that the spinner is more likely to stop in the bigger sectors. Or they count two-dice sums and find that seven is more likely to occur than 12. Or they flip coins and discover that, after three heads in a row, heads is still as likely to come up as tails.

Units about games and rules help elementary students develop four important mathematical ideas: elementary understanding of probability; the correspondence between rules, games, instructions, and actions; the concept of a strategy; and the mathematical bases of fairness. Experience with games

prepares students for many topics in discrete mathematics as well as logic and language.

Sharing. Sharing often involves equal portions and thus can involve notions of multiplication, division, fractions, fairness, and average. How can a jar of candy be shared by the class? How can ten cookies be shared by four people? How can a pizza be shared? How the work for the project be shared? How much time should each person work in the booth at the carnival? What is the average number of buttons worn by students in the class? (What is a fair share of buttons?)

Sometimes a single object is shared with others; or the number of objects to be shared is greater than the number of people; or the number of objects to be shared is less than the number of people. Often, the sharing does not come out even. What can be done with the leftovers?

Young children are used to sharing their toys with each other. They also take turns, share experiences orally, and share treats. Ideas about fair share come up when, for example, students try to fill everyone's juice cup to the same level. (How much juice is needed in the pitcher so that everyone may have a cup?) When young students share many objects, they generally distribute them one by one. When they share one object, or when the number of objects is fewer than the number of people, they break the objects into smaller pieces.

As students have more experiences, they see how their sharing can be represented by division and fractions. Their work in sharing prepares them for the mathematics of equality, proportion, rational and real numbers.

Content of the Middle School Program

The mathematical content for the middle grades differs from the content for the elementary grades in the level of abstraction and variety of representations. In the middle grades students encounter new concepts in anticipation of the abstractions of the secondary program. They also revisit mathematical ideas introduced in the earlier grades, although at a more abstract level. Students move from number to variable, from specifics to generalization, and from description to informal proof in the middle grades. Although middle grade students continue to refine their number sense through meaningful interactions with estimation, mental arithmetic, and judgments about the reasonableness of results, they also begin to develop a broader mathematical sense beyond the realm of numbers. They need to experience mathematical ideas in a way that gives the students a genuine appreciation of the depth of those ideas and not necessarily the formal vocabulary or rules.

The content that follows is important and appropriate for all students. Some may develop greater depth and sophistication, but all can and must develop a deep understanding of all the unifying ideas. (See "Moving Away from Tracking," pages 62–66.) Thus, this content will be the basis for strong courses (called here Math 6, Math 7, and Math 8) that prepare many students for the three high school core courses. The general mathematics courses that dwell on computational facility should be eliminated. (See pages 130–36.)

"Middle school mathematics should emphasize the practical power of mathematics.

If instruction is to give students mathematical power, then problem solving needs to be emphasized throughout all grades. Students need to perceive mathematics as more than the subject matter itself—as, in fact, a discipline of reasoning that enables them to attack and solve problems of increasing difficulty and complexity. A focus on problems rather than just exercises is important throughout the curriculum.

Broadening the elementary school curriculum has important implications about entry into secondary mathematics. The middle grades should not be viewed as a time for consolidation or as a pause for a rest, but as an essential part of a child's mathematical development. Its focus should be on mathematics for everyday life, a theme rich in motivation that leads naturally to many important mathematical topics (e.g., data analysis, geometric measurement, interest rates, and spreadsheet analysis). Understanding the concepts of elementary school mathematics is essential for the study of secondary school mathematics; however, proficiency in the procedures of hand arithmetic computation should no longer be the critical factor in judging student readiness for advanced study."

—*Reshaping School Mathematics*, p. 44

The NCTM standards for the middle grades are excerpted below. By the end of the eighth grade, students should be able to do what is described in the standards. Those concerned with grade four and below may wish to consult the kindergarten through grade four standards, which appear on pages 105–7. These excerpts are not intended to be self-explanatory; therefore, the reader is advised to see the full discussion in the NCTM's *Curriculum and Evaluation Standards for School Mathematics.* (See also Appendix E in this *Framework.*)

Standard 1: Mathematics as Problem Solving

In grades five through eight, the mathematics curriculum should include numerous and varied experiences with problem solving as a method of inquiry and application so that students can:

- Use problem-solving approaches to investigate and understand mathematical content.
- Formulate problems from situations within and outside mathematics.
- Develop and apply a variety of strategies to solve problems, with emphasis on multistep and nonroutine problems.
- Verify and interpret results with respect to the original problem situation.
- Generalize solutions and strategies to new problem situations.
- Acquire confidence in using mathematics meaningfully.

Standard 2: Mathematics as Communication

In grades five through eight, the study of mathematics should include opportunities to communicate so that students can:

- Model situations, using oral, written, concrete, pictorial, graphical, and algebraic methods.
- Reflect on and clarify their own thinking about mathematical ideas and situations.
- Develop common understandings of mathematical ideas, including the role of definitions.
- Use the skills of reading, listening, and viewing to interpret and evaluate mathematical ideas.
- Discuss mathematical ideas and make conjectures and convincing arguments.
- Appreciate the value of mathematical notation and its role in the development of mathematical ideas.

Standard 3: Mathematics as Reasoning

In grades five through eight, reasoning should permeate the mathematics curriculum so that students can:

- Recognize and apply deductive and inductive reasoning.
- Understand and apply reasoning processes, with special attention being given to spatial reasoning and reasoning with proportions and graphs.
- Make and evaluate mathematical conjectures and arguments.

- Validate their own thinking.
- Appreciate the pervasive use and power of reasoning as a part of mathematics.

Standard 4: Mathematical Connections

In grades five through eight, the mathematics curriculum should include the investigation of mathematical connections so that students can:

- See mathematics as an integrated whole.
- Explore problems and describe results, using graphical, numerical, physical, algebraic, and verbal mathematical models or representations.
- Use a mathematical idea to further their understanding of other mathematical ideas.
- Apply mathematical thinking and modeling to solve problems that arise in other disciplines, such as art, music, psychology, science, and business.
- Value the role of mathematics in our culture and society.

Standard 5: Number and Number Relationships

In grades five through eight, the mathematics curriculum should include the continued development of number and number relationships so that students can:

- Understand, represent, and use numbers in a variety of equivalent forms (integer, fraction, decimal, percent, exponential, and scientific notation) in real-world and mathematical problem situations.
- Develop number sense for whole numbers, fractions, decimals, integers, and rational numbers.

- Understand and apply ratios, proportions, and percents in a wide variety of situations.
- Investigate relationships among fractions, decimals, and percents.
- Represent numerical relationships in one-dimensional and two-dimensional graphs.

Standard 6: Number Systems and Number Theory

In grades five through eight, the mathematics curriculum should include the study of number systems and number theory so that students can:

- Understand and appreciate the need for numbers beyond the whole numbers.
- Develop and use order relations for whole numbers, fractions, decimals, integers, and rational numbers.
- Extend their understanding of whole number operations to fractions, decimals, integers, and rational numbers.
- Understand how the basic arithmetic operations are related to one another.
- Develop and apply number theory concepts (e.g., primes, factors, and multiples) in real-world and mathematical problem situations.

Standard 7: Computation and Estimation

In grades five through eight, the mathematics curriculum should develop the concepts underlying computation and estimation in various contexts so that students can:

- Compute with whole numbers, fractions, decimals, integers, and rational numbers.
- Develop, analyze, and explain procedures for computation and techniques for estimation.

- Develop, analyze, and explain methods for solving proportions.
- Select and use an appropriate method for computing from among mental arithmetic, paper-and-pencil, calculator, and computer methods.
- Use computation, estimation, and proportions to solve problems.
- Use estimation to check the reasonableness of results.

Standard 8: Patterns and Functions

In grades five through eight, the mathematics curriculum should include explorations of patterns and functions so that students can:

- Describe, extend, analyze, and create a wide variety of patterns.
- Describe and represent relationships with tables, graphs, and rules.
- Analyze functional relationships to explain how a change in one quantity results in a change in another.
- Use patterns and functions to represent and solve problems.

Standard 9: Algebra

In grades five through eight, the mathematics curriculum should include explorations of algebraic concepts and processes so that students can:

- Understand the concepts of variable, expression, and equation.
- Represent situations and number patterns with tables, graphs, verbal rules, and equations and explore the interrelationships of these representations.
- Analyze tables and graphs to identify properties and relationships.

- Develop confidence in solving linear equations, using concrete, informal, and formal methods.
- Investigate inequalities and nonlinear equations informally.
- Apply algebraic methods to solve a variety of real-world and mathematical problems.

Standard 10: Statistics

In grades five through eight, the mathematics curriculum should include exploration of statistics in real-world situations so that students can:

- Systematically collect, organize, and describe data.
- Construct, read, and interpret tables, charts, and graphs.
- Make inferences and convincing arguments based on data analysis.
- Evaluate arguments based on data analysis.
- Develop an appreciation for statistical methods as powerful means for decision making.

Standard 11: Probability

In grades five through eight, the mathematics curriculum should include explorations of probability in real-world situations so that students can:

- Model situations by devising and carrying out experiments or simulations to determine probabilities.
- Model situations by constructing a sample space to determine probabilities.
- Appreciate the power of using a probability model by comparing experimental results with mathematical expectations.
- Make predictions based on experimental or theoretical probabilities.

120

- Develop an appreciation for the pervasive use of probability in the real world.

Standard 12: Geometry

In grades five through eight, the mathematics curriculum should include the study of the geometry of one, two, and three dimensions in a variety of situations so that students can:

- Identify, describe, compare, and classify geometric figures.
- Visualize and represent geometric figures, with special attention to developing spatial sense.
- Explore transformations of geometric figures.
- Represent and solve problems using geometric models.
- Understand and apply geometric properties and relationships.
- Develop an appreciation of geometry as a means of describing the physical world.

Standard 13: Measurement

In grades five through eight, the mathematics curriculum should include extensive concrete experiences using measurement so that students can:

- Extend their understanding of the process of measurement.
- Estimate, make, and use measurements to describe and compare phenomena.
- Select appropriate units and tools to measure to the degree of accuracy required in a particular situation.
- Understand the structure and use of systems of measurement.
- Extend their understanding of the concepts of perimeter, area, volume, angle measure, capacity, and weight and mass.
- Develop the concepts of rates and other derived and indirect measurements.
- Develop formulas and procedures for determining measures to solve problems.

Regarding the following outline of content, it should be remembered that brief labels do not adequately convey the subject. The breadth of the middle grade curriculum is specified in the strands. The NCTM standards are incorporated in this *Framework* as desired outcomes of the middle school program. By the end of the new course Math 8, students should be able to do what is described in the standards.

Strands

Functions

Algebra

Geometry

Statistics and Probability

Discrete Mathematics

Measurement

Number

Logic and Language

Unifying Ideas

Proportional Relationships

Multiple Representations

Patterns and Generalization

Suggested Units for the Middle Grades

Objects, Shapes, and Containers

Maps and Scale Drawings

Growth

Motion

Expressing Proportional Relationships

Unifying Ideas for the Middle Grades

Much of the mathematical content in the middle grades can be organized around three unifying ideas: *proportional relationships*, *multiple representations*, and *patterns and generalization*.

These are not specific subjects to be covered only at particular times; rather, they are themes that are present continually, integrating the different things students are learning. Further, each theme is specifically meant to be developed as a bridge to help students make the transition from elementary school mathematics to high school mathematics.

Proportional relationships. Proportional relationships are at the heart of the most basic quantitative understanding of the world and are continually useful in school, in the workplace, and in everyday life. Students in the middle grades should explore a variety of situations in depth in which proportional relationships play a major role and work to develop a broad sense of the kind of proportional reasoning that applies in these situations.[1]

The central concept is the representation of one quantity as a certain *proportion* of another. This concept plays a key role in many basic subjects often studied in middle school: ratio, rate, "per," percent quantities and per-unit quantities, proportional parts, slope, similarity, scale relationships, linear functions, and probability. Although these subjects should be developed separately, a considerable amount of time should be devoted to illustrating how they all involve a kind of proportional relationship. For example, in a sloped straight line, the *rise* is proportional to the *run*, and their ratio is the slope of the line; in a scale drawing real distances are proportional to distances in the drawing; in a class of similar figures, any two lengths in one figure are in the same proportion as the corresponding two lengths in any other figure; and when an event has a certain probability, the event happens a certain fixed proportion of the time. Only by making these kinds of connections will students begin to see the individual things they are learning as parts of a bigger idea rather than as isolated fragments.

When students complete eighth grade, they should have a solid background of experiences with similarity and simple linear functions. The idea of proportional scaling (enlarging and shrinking by a scale factor), which connects similarity and linearity, is especially important.

Multiple representations. Mathematical representations are ways of presenting information and relationships that arise when problems are analyzed. They include a broad selection of devices: sketches, orthogonal and isometric views of objects,

---

[1]The idea of proportional relationships is intended here in a sense considerably broader than the topics *ratio, proportion,* and *percent* as they typically appear in middle school textbooks.

schematic diagrams, tables, frequency distributions, charts, graphs, physical models, and prose exposition. They also include representations made with the use of a computer, such as graphs, graphic displays, simulations, and spreadsheet analyses. Each type of representation illuminates different aspects of a relationship. Students should routinely construct *multiple* representations of problems they are working on.

Mathematical representations are powerful tools for visualizing and understanding problem situations, communicating mathematically, and solving problems. Too frequently, students are exposed only to a narrow range of mathematical representations centered on numbers and literal symbols for numbers. They need to see that it is easier to understand important mathematical ideas when these ideas are represented in several ways and that different representations bring out different aspects of the idea. Moreover, for individual students, specific kinds of representations may initially clarify the ideas more effectively than others.

Some subject matter related to representation is already standard fare. For example, *coordinate systems* are a major vehicle of mathematical representation. Students will work with a variety of one-dimensional and two–dimensional coordinate systems to represent functional relationships, geometric shapes, and statistical information.

In another example students continue to work with appropriate forms for representing numbers: as decimals, as fractions, as percent or per-unit quantities, or in scientific notation. During their middle school years, students will work toward effective use of variables and variable expressions as powerful representational tools and will build the foundation for understanding formulas, functions, and equations.

Recent emphasis on *spatial visualization* presents another important aspect of representation. If students are expected to carry out meaningful work on genuine problems, they must be able to visualize these problems effectively and work with specific representations. Models and computer graphics can be valuable here.

Many aspects of the idea of multiple representation are not treated nearly well enough in typical programs. Mathematically powerful students should be able to produce a revealing sketch of a problem or know how to use a geometric analogy to shed light on a nongeometric problem. Moreover, it is not sufficient to illustrate specific methods of representation one at a time.

These kinds of representation need to be used *together* to provide complete and balanced mathematical analyses.

Patterns and generalization. Mathematics has been characterized as the science of patterns. For situations to be analyzed effectively, patterns must be observed, generalizations made, and mathematics used to illuminate the features of the situation. In elementary school, students learn to look for, describe, and create simple patterns involving numbers and shapes. In the middle school years, they extend this understanding as they confront more complex and varied patterns, progressing from recognizing simple numerical and geometric patterns. The next steps are becoming able to describe many kinds of regularities of number or shape and making, testing, and then utilizing generalizations about given information. Useful analysis of real situations involves the ability to see patterns, make generalizations, and manipulate the situations to the point at which the mathematics can illuminate features of the original situation.

Some specific ways in which students extend their understanding of patterns in middle school are to use (1) a variable expression—to generalize a numerical pattern; (2) the idea of similarity—to capture what is alike in a set of figures that have the same shape but differ in size; (3) different kinds of symmetry—to describe repeating geometric patterns; (4) systematic patterned lists of arrangements as a way of enumerating possible configurations of objects or choices; and (5) ideas from probability—to express statistical patterns in data.

By the end of the eighth grade, students should be able to express many of their generalizations, using words, algebraic symbols, and geometric diagrams.

Suggested Units for the Middle Grades

The following are offered as possible ways for organizing mathematics into coherent units for the middle grade curriculum. Remember that this set is not complete. Other units might have very different characteristics: different settings for the mathematics, different instructional styles, and different types of student projects.

Objects, shapes, and containers. In middle school, students spend considerable time studying three-dimensional objects—both the idealized shapes of solid geometry and the way in which these shapes are illustrated by models, drawings, videos, and computer graphics. Much of the focus is placed on a few

"When viewed in [a] broader context, we see that mathematics is not just about number and shape, but about pattern and order of all sorts. Number and shape—arithmetic and geometry—are but two of many media in which mathematicians work. Active mathematicians seek patterns wherever they arise."

—*On the Shoulders of Giants*, p. 2

basic three-dimensional shapes: *rectilinear solids*, *prisms*, *cylinders*, *square base pyramids*, and *cones*. These shapes can be viewed both as *solids* and also as hollow *containers* that can be used to measure the volume of liquids. Students also study more complex natural shapes, using decomposition and estimation to relate knowledge of basic shapes to natural shapes.

As an outcome of this work, students develop specific abilities. They can *visualize* three-dimensional shapes; *represent* the shapes with sketches, orthogonal views, isometric views, and plane sections; and *communicate* information about shapes in verbal descriptions, using the concepts of line, plane, angle, perpendicularity, and parallelism. In addition, they discuss the *size* of basic shapes, using one-dimensional measures such as *width* and *girth*, two-dimensional measures such as *surface area* and *footprint area*, and three-dimensional measures such as *volume* and *capacity*. They use standard *formulas* for these size characteristics and construct formulas of their own when they consider new variations or combinations of these shapes. Finally, they are comfortable in going beyond the purely geometric measures of length, area, and volume to discuss objects in terms of their material characteristics of *weight* and *density*.

Maps and scale drawings. Maps and scale drawings are ways to represent on paper the features of real objects or places. A scale drawing is often a *reduced* version of something; but when more detail needs to be seen in a tiny object, it may be a *blown-up* version.

Scale involves intuitively powerful and mathematically precise forms of proportional reasoning; maps and scale drawings call for multiple representations of geometric and numerical objects.

The central concept—the *scale* of a map or drawing—can be indicated either as an equivalence (such as 1 cm = 2.5 m) or as a ratio (such as 1 to 250). Students explore how a change in scale affects the measures of *lengths*, *areas*, and *angles*. Because every *distance* on a drawing is proportional to the corresponding real distance, the constant of proportionality for distance is the same as the scale of the drawing.

Every *angle* in a drawing is the *same* as the corresponding real angle, and the *shapes* in the drawing are the same as the real shapes. Distortions in shape arise only in special circumstances—for example, if the vertical direction has a different scale than the horizontal scale.

Students read and construct many different kinds of maps and scale drawings. They become familiar with the idea of *similar figures*—figures that have the same *shape* but perhaps differ in size. In particular, they learn the importance of the idea of *similar right triangles* in analyzing the relationship of the scale drawing to the full-size entity.

As part of their work with maps, students explore the idea of a *path* and learn to specify straight line paths by giving instructions for *distance* (the length of a straight line segment to be traversed) and *direction* (the angle of turn at the end of each segment). Students can explore paths both in the field (through orienteering activities) and on paper. Similarly, they can approach paths both graphically and computationally. In the latter case the Pythagorean theorem and facts about 30-60-90 and 45-45-90 triangles are basic and central.

Growth. Students explore the growth of several kinds of things—plants, animals, themselves, populations, investments, and debts. Because the primary emphasis is placed on linear growth (to follow the unifying idea of proportionality), students construct and analyze many straight-line functions and graphs of size versus time.

For contrast, students also need to work with situations involving nonlinear growth. For example, students can be asked to explore two ways in which the number *12* can grow to the number *20* in four steps. In one way the same number is *added* at each step; in the other way the same number is *multiplied* at each step. Working with a calculator, students can gain valuable intuition about the difference between additive (linear) growth and multiplicative (exponential) growth.

In another example the growth of a circle (exemplified perhaps by the cross section of a tree trunk) can be represented in ten snapshots taken at equal time intervals. Students can contrast three different ways the circle can grow. In one, the *radius* increases by a constant amount in each time interval; in another, an equal *area* is added to the circle in each time interval; and in a third, the radius increases by a constant *factor* in each time interval. This kind of analysis can be carried out in conjunction with a study of the actual growth of trees.

Motion. To understand the concept of motion and how it can be represented mathematically, students explore motion by timing their own walking, running, or biking; studying bus or train timetables; investigating moving sidewalks or escalators; or timing objects as they roll down inclined planes.

The relationship of three kinds of quantities needs to be understood: a geometric quantity (position or distance); a temporal quantity (time or elapsed time interval); and a *rate* that relates the two (speed or velocity). In all the work with motion, students constantly use *graphs* to give a visual representation of the situation. The most concrete kind of graph is one that shows the position of one or more objects moving along a path as a function of the clock time; such a graph clearly shows meeting (or collision) points. Other graphs might show distance traveled as a function of time intervals; such graphs, all going through the origin, can be used to compare speeds. Finally, speed itself can be graphed as a function of time.

Students devote considerable time to analyzing motion at a constant speed and its corresponding straight line graphs. Students can analyze motion quantitatively in terms of the underlying relationship (distance traveled) = (speed) × (elapsed time); but emphasis is placed on *using* this relationship in a context to arrive at meaningful results rather than on simply plugging in given numbers.

Expressing proportional relationships. After students have substantial experience with proportional relationships embedded in intuitively meaningful situations, they study ways to express such relationships as functions. They learn to use functional expressions to investigate and compare relationships with which they are familiar. There is work with a basic library of functions: pure linear functions of the form $y = kx$ that express proportional relationships between y and x; offset linear functions of the form $y = y_0 + kx$ that express proportional relationships between $(y - y_0)$ and x; and functions of the form $y = \frac{k}{x}$ that express inverse proportions.

Reflecting on earlier investigations, students revise earlier work on proportions based on new understanding.

The examples previously given are far from sufficient. Curriculum developers will need to develop units that give students appropriate experiences with the range of issues described in the standards, emphasizing unifying ideas. These might include units cohering around interpretation of data; fairness; meters and gauges (as decimal representations); reflection on the operations of arithmetic—their meanings and the connections among them; units of measurement; and descriptive geometry.

Mathematical Content and Course Structure in Grades Nine Through Twelve

C h a p t e r 5

This chapter contains a description of the specific mathematical content and course structure for grades nine through twelve. First, the high school core sequence is discussed, and then the question of content is addressed. Next, the standards developed by The National Council of Teachers of Mathematics for grades nine through twelve are presented and are followed by the strands and unifying ideas for those grade levels. Finally, subject matter for potential units of instruction is suggested.

High school mathematics has traditionally acted as a filter. That is, a student who does not progress through high school mathematics suffers a loss of job skills, citizenship skills, and access to many colleges and universities. The current rate of students being filtered out each year in the sequence is an unacceptable failure—not of students but of the program. That so many students do not complete the program confounds the issues of expectations for student performance and standards for mathematical content. Specifying high standards only for students merely filters out more students. High standards must also be set for the program, and what the program is to accomplish for every student must be specified.

Not just the students who drop out of the traditional sequence in mathematics are ill-served. Too many of the students who successfully progress through typical college preparatory mathematics courses later show up in college classrooms with disappointing levels of understanding and negative dispositions toward mathematics. This situation must also be changed.

To address these problems, this *Framework* describes a common core sequence for all high school students; that is, three years of sound mathematical foundation for citizenship, for the workplace, and for further study.

High School Core Sequence

This *Framework* endorses the recommendations contained in NCTM's *Curriculum and Evaluation Standards for School Mathematics* for a core high school curriculum for *all* students. It further recommends that all students progress through a common core sequence consisting of three courses: Course 1, Course 2, and Course 3. Each course will have content drawn from all the strands so that students will learn the content described on pages 141–63. However, the specific configuration of this content within the courses will not be specified in this *Framework*.

All high school students can complete all three core courses, although the students may begin at different times. As the accompanying figure suggests, the typical student will enter the common course sequence in the ninth grade. Students needing transitional work will take Math A with the intention of entering

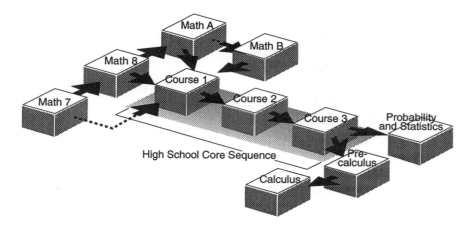

Math A

Math B

Math 8

Course 1

Math 7

Course 2

Course 3

Probability and Statistics

High School Core Sequence

Pre-calculus

Calculus

the sequence the following year. At the end of the seventh grade, exceptional students may take Course 1 as eighth graders. Math B will be offered only for students who have experienced serious difficulties in Math A. Ideally, Math B will become unnecessary after a time, and all secondary students will complete the full three years of the core sequence.

Because all students should study mathematics every year, attractive mathematics courses must be created at each level. Students deserve programs worthy of four years of study. Programs consistent with this *Framework* should consist of courses that encourage all students to develop a positive attitude toward mathematics. Such courses should be rigorous and challenging as well as engaging and friendly. The instructional materials, the teaching practices, and assessment should focus not on screening for talent but on developing everybody's mathematical power.

Developing Mathematical Power in the Core Sequence

The high school core program will have all of the attributes described elsewhere in this *Framework*, including, for example, increased student writing; a changed role for the teacher; more holistic assessment practices; work in heterogeneous groups; and large projects. The curriculum should not be organized by strand but should rather be integrated across strands in coherent units that emphasize a deepening understanding of unifying ideas. As described in Chapter 3, these units provide ample opportunities for student work that demonstrates all of the dimensions of mathematical power. Students' understanding should deepen across the curriculum from unit to unit and from year to year.

"The program will require a shift in emphasis from a curriculum dominated by memorization of isolated facts and procedures to one that emphasizes conceptual understandings, multiple representation and connection, mathematical modeling, and mathematical problem solving."

—*Curriculum and Evaluation Standards for School Mathematics*, p. 125

"All students should study mathematics every year they are in school."

—*Everybody Counts*, p. 50

131

By their nature high school courses in algebra, geometry, and Algebra II have divided the curriculum and have confined student development to specific areas. How can the development of various topics in mathematics in the core sequence overcome these disadvantages? Some examples follow:

1. *Geometry*

   Geometry should receive considerable attention in all three courses from two perspectives. First, from a more synthetic perspective, there can be a focus on shapes growing out of student experiences with objects. In Course 1 the focus might be on similarity, which connects to linearity and scale factor. Second, from a more algebraic perspective, there can be a focus on space or location growing out of student experiences with location and distance.

   For example, in Course 1 a focused study of *distance* in coordinate space based on the Pythagorean theorem can lead to work with triangles as the basis for remote measurement. (This approach refers back to similarity and direct and indirect variation.) Course 2 can develop ideas associated with *direction:* angle, parallels, and direction in coordinate space. In Course 3 the unit circle can tie many of these ideas together as the basis for a more general understanding of number and functions, including trigonometry. In all courses a considerable amount of work should be done with three-dimensional objects and spaces of one, two, and three dimensions.

2. *Functions*

   The study of variation and functions can dwell in Course 1 on in-depth work with direct and indirect variation and exponential growth in a variety of realistic situations that provide meaning. Scale factor, slope, and linearity should be developed in this context. In Course 2 quadratics can initiate the study of polynomial functions and their applications rather than the assortment of unmotivated manipulations of quadratics found in typical Algebra I courses.

   As students use a growing variety of functions over the three years, they should learn to compare the behavior of different kinds of functions, using graphing calculators and computers to do so. This comparative view of functions should include developing the ability to see the relationships that make one or another kind of function useful for making sense of a given situation. (Exponential functions are often used to understand situations related to growth. Why?) Course 3 should include a look back over the range of situations studied and the manner in which different functions were used.

3. *Symbol Manipulation*

Development of the student's facility with the symbolic language of mathematics and graphical representations should receive attention from the beginning in Course 1. Emphasis should be placed at once on work in concrete situations that is supplemented with the explicit development of craft and technique. As students progress through Course 2 and Course 3, they should develop increasing facility in manipulating symbols and rendering concrete situations symbolically. By the end of Course 3, students should be able to use symbolic and graphical representations to investigate abstract situations and problems.

Options After the Core Sequence

All students should complete the core curriculum. Afterwards, students with differing goals and interests should have options. College-preparatory students should take mathematics in every year of high school. Many of these students can and should proceed through a course in mathematical analysis followed by advanced-placement calculus AB or BC. (See Appendix D for a statement on calculus by the Mathematics Association of America [MAA] and the National Council of Teachers of Mathematics [NCTM].) Many other college-preparatory students and other students would be better served by post-core courses that treat selected material from the post-core content in an applications-based format.

When calculus is offered in grade twelve, the current recommended practice for most schools is to follow one of the two syllabi of the Advanced Placement Program of the College Board. In accordance with the MAA and NCTM recommendation, all students completing the course should expect to take the appropriate advanced placement examination as an integral important part of the course. If advanced placement courses are offered, sufficient resources should be provided to enable all students enrolled to take the advanced-placement examinations.

After the core curriculum has been completed, even college-preparatory students need an alternative to moving toward calculus as quickly as possible. Many students would be better served in grades eleven or twelve with a course in probability and statistics that emphasizes application to problems encountered in ordinary living as well as applications to history–social science and the life and physical sciences. This course would be available to students completing the three-year core sequence and to students who had taken the mathematical analysis or precalculus course but would prefer to delay calculus until the first year in college. Schools may elect to change the emphasis in different sections of Course 3, depending on whether students expect to take precalculus or an alternative mathematics course the following year, although the instructional practices and resources would be the same, regardless of emphasis.

Students with vocational goals that do not include attending a four-year college can benefit from mathematics taught in an integrated vocational course. (However, see pages 101–2 of this *Framework* about pitfalls to avoid when integrating mathematics with other subjects.) The mathematical content of integrated courses should meet standards for the core program.

The goal of this *Framework* is to achieve a core sequence and associated courses that challenge the best students and are accessible to the vast majority. If this goal is to be achieved by the year 2000, the first steps must be taken now. Time and support will be needed for teachers and schools to make the required changes; and, as the changes occur, more and more students will succeed. How far and how fast students can go remains to be demonstrated.

Difficulties in Implementing the Core Sequence

The three-year core sequence represents a major restructuring of high school mathematics education. Schools and school districts will need to be creative and diligent to overcome formidable political and logistical problems, some of the most serious being acknowledged. Nevertheless, a start must be made. Continuing dialogue and open communication will help in the sharing of solutions and coping with inevitable difficulties.

Transfers between programs. Because this *Framework* does not specify the content for the three courses separately, what will happen when students transfer between core courses at different schools or between traditional courses and newly implemented integrated programs? Several things should be kept in mind when considering this matter. First, the problem is not special to this program. Algebra II typically begins with weeks of review because students have transferred in from traditional geometry classes in which little time was spent on *algebra* and even less on *functions*. Transferring is always difficult. Second, heterogeneous groups and peer help provide a safety net to help new students catch up. Third, self-contained units that support the same unifying ideas allow students a relatively fresh start every six weeks or so, with recurring major ideas.

Need for instructional materials. To implement the core sequence without instructional materials is difficult. Although suitable materials are not yet widely available, hundreds of California teachers and university faculty are involved in projects to develop integrated high school curricula. These projects are creating and offering new courses designed for a wider span of students. Many recent textbooks from commercial publishers are also moving toward more integrated versions of the traditional sequence. As these materials become available, implementation of the core sequence will become feasible in a wider range of communities.

Curriculum designers will continue to develop materials for integrated programs—materials that encompass the content of the NCTM standards without dividing it along traditional boundaries. At the same time mathematics departments and teachers are encouraged to take whatever steps toward the integrated core sequence that are practical for them and their students.

Need for technology. As discussed in Chapter 2 (page 59), students should be provided with regular, systematic access to display calculators with graphing capabilities—or to technology that supersedes them (e.g., notebook microcomputers)—beginning in Course 1. Student learning of mathematical concepts can benefit from the insights that high-speed, accurate computation affords. A technology-rich environment can become a central part of the learning experience, not just an enhancement to time-honored but often outdated paper-and-pencil mathematical methods. But technology requires money for purchase and enough instructional time so that students and teachers are comfortable. The course must be about mathematics, not about learning to use a device.

Student preparation and support. The core sequence should be as rigorous as that in any high-quality college-preparation program. Accordingly, homework and significant study outside the class are required. Many students will need an organized support structure beyond the classroom for studying difficult material—a structure that takes time and resources. In addition, a greater number of students may need to take Math A and Math B in the first few years of restructuring to prepare for the kind of mathematics included in the core courses. And as courses in the lower grades are improved, all students should be expected to participate fully.

The core sequence embraces the explicit expectation that students will not be assigned to different sections according to imagined or measured ability in mathematics or according to their intention to attend college. Although the immediate detracking of most mathematics courses may not be desirable or even possible, the goal remains constant. Practical ways must be found to make this vision a reality.

NCTM Standards for High School

Now that the core sequence has been discussed, it is appropriate to examine the mathematical content of its courses. One source of guidance is the NCTM standards for high school that are excerpted here. By the end of the twelfth grade, students should be able to do what is described in these standards. The following excerpts are not intended to be self-explanatory; therefore, the reader is advised to see the full discussion in the *Curriculum and Evaluation Standards for School Mathematics* published by NCTM. (See also Appendix E in this *Framework*.)

After the standards come a description of the strands and unifying ideas for high school and suggestions as to what mathematics might be addressed in coherent high school units of instruction.

Standard 1: Mathematics as Problem Solving

In grades nine through twelve, the mathematics curriculum should include the refinement and extension of methods of mathematical problem solving so that all students can:

- Use, with increasing confidence, problem-solving approaches to investigating and understanding mathematical content.
- Apply integrated mathematical problem-solving strategies to solve problems from within and outside mathematics.
- Recognize and formulate problems from situations within and outside mathematics.
- Apply the process of mathematical modeling to real-world problem situations.

Standard 2: Mathematics as Communication

In grades nine through twelve, the mathematics curriculum should include the continued development of language and symbolism to communicate mathematical ideas so that all students can:

- Reflect on and clarify their thinking about mathematical ideas and relationships.
- Formulate mathematical definitions and express generalizations discovered through investigations.
- Express mathematical ideas orally and in writing.
- Read written presentations of mathematics with understanding.
- Ask clarifying and extending questions related to mathematics they have read or heard about.
- Appreciate the economy, power, and elegance of mathematical notation and its role in the development of mathematical ideas.

Standard 3: Mathematics as Reasoning

In grades nine through twelve, the mathematics curriculum should include numerous and varied experiences that reinforce and extend logical reasoning skills so that all students can:

- Make and test conjectures.
- Formulate counterexamples.
- Follow logical arguments.
- Judge the validity of arguments.
- Construct simple valid arguments.

And so that, in addition, planning to attend college students can:

- Construct proofs for mathematical assertions, including indirect proofs and proofs by mathematical induction.

Standard 4: Mathematical Connections

In grades nine through twelve, the mathematics curriculum should

include investigation of the connections and interplay among various mathematical topics and their applications so that all students can:

- Recognize equivalent representations of the same concept.
- Relate procedures in one representation to procedures in an equivalent representation.
- Use and value the connections among mathematical topics.
- Use and value the connections between mathematics and other disciplines.

Standard 5: Algebra

In grades nine through twelve, the mathematics curriculum should include the continued study of algebraic concepts and methods so that all students can:

- Represent situations that involve variable quantities with expressions, equations, inequalities, and matrices.
- Use tables and graphs as tools to interpret expressions, equations, and inequalities.
- Operate on expressions and matrices and solve equations and inequalities.
- Appreciate the power of mathematical abstraction and symbolism.

And so that, in addition, students planning to attend college can:

- Use matrices to solve linear systems.
- Demonstrate technical facility with algebraic transformations, including techniques based on the theory of equations.

Standard 6: Functions

In grades nine through twelve, the mathematics curriculum should include the continued study of functions so that all students can:

- Model real-world phenomena with a variety of functions.

- Represent and analyze relationships, using tables, verbal rules, equations, and graphs.
- Translate among tabular, symbolic, and graphical representations of functions.
- Recognize that a variety of problem situations can be modeled by the same type of function.
- Analyze the effects of parameter changes on the graphs of functions.

And so that, in addition, students planning to attend college can:

- Understand operations on, and the general properties and behavior of, classes of functions.

Standard 7: Geometry from a Synthetic Perspective

In grades nine through twelve, the mathematics curriculum should include the continued study of the geometry of two and three dimensions so that all students can:

- Interpret and draw three-dimensional objects.
- Represent problem situations with geometric models and apply properties of figures.
- Classify figures in terms of congruence and similarity and apply these relationships.
- Deduce properties of, and relationships between, figures from given assumptions.

And so that, in addition, students planning to attend college can:

- Develop an understanding of an axiomatic system through investigating and comparing various geometries.
- Use, with increasing confidence, problem-solving approaches to investigate and understand mathematical content.

- Apply integrated mathematical problem-solving strategies to solve problems from within and outside mathematics.
- Recognize and formulate problems from situations within and outside mathematics.
- Apply the process of mathematical modeling to real-world problem situations.

Standard 8: Geometry from an Algebraic Perspective

In grades nine through twelve, the mathematics curriculum should include the study of the geometry of two and three dimensions from an algebraic point of view so that all students can:

- Translate between synthetic and coordinate representations.
- Deduce properties of figures, using transformations and using coordinates.
- Identify congruent and similar figures, using transformations.
- Analyze properties of Euclidean transformations and relate translations to vectors.

And so that, in addition, students planning to attend college can:

- Deduce properties of figures using vectors.
- Apply transformations, coordinates, and vectors in problem solving.

Standard 9: Trigonometry

In grades nine through twelve, the mathematics curriculum should include the study of trigonometry so that all students can:

- Apply trigonometry to problem situations involving triangles.
- Explore periodic real-world phenomena, using the sine and cosine functions.

And so that, in addition, students planning to attend college can:

- Understand the connection between trigonometric and circular functions.
- Use circular functions to model periodic real-world phenomena.
- Apply general graphing techniques to trigonometric functions.
- Solve trigonometric equations and verify trigonometric identities.
- Understand the connections between trigonometric functions and polar coordinates, complex numbers, and series.

Standard 10: Statistics

In grades nine through twelve, the mathematics curriculum should include the continued study of data analysis and statistics so that all students can:

- Construct and draw inferences from charts, tables, and graphs that summarize data from real-world situations.
- Use curve fitting to predict from data.
- Understand and apply measures of central tendency, variability, and correlation.
- Understand sampling and recognize its role in statistical claims.
- Design a statistical experiment to study a problem, conduct the experiment, and interpret and communicate the outcomes.
- Analyze the effects of data transformations on measures of central tendency and variability.

And so that, in addition, students planning to attend college can:

- Transform data to aid in data interpretation and prediction.
- Test hypotheses, using appropriate statistics.

Standard 11: Probability

In grades nine through twelve, the mathematics curriculum should include the continued study of probability so that all students can:

- Use experimental or theoretical probability, as appropriate, to represent and solve problems involving uncertainty.
- Use simulations to estimate probabilities.
- Understand the concept of a random variable.
- Create and interpret discrete probability distributions.
- Describe, in general terms, the normal curve and use its properties to answer questions about sets of data that are assumed to be normally distributed.

And so that, in addition, students planning to attend college can:

- Apply the concept of a random variable to generate and interpret probability distributions, including binomial, uniform, normal, and chi square.

Standard 12: Discrete Mathematics

In grades nine through twelve, the mathematics curriculum should include topics from discrete mathematics so that all students can:

- Represent problem situations, using discrete structures such as finite graphs, matrices, sequences, and recurrence relations.
- Represent and analyze finite graphs, using matrices.

139

- Develop and analyze algorithms.
- Solve enumeration and finite probability problems.

And so that, in addition, students planning to attend college can:

- Represent and solve problems, using linear programming and difference equations.
- Investigate problem situations that arise in connection with computer validation and the application of algorithms.

Standard 13: Conceptual Underpinnings of Calculus

In grades nine through twelve, the mathematics curriculum should include the informal exploration of calculus concepts from both a graphical and a numerical perspective so that all students can:

- Determine maximum and minimum points of a graph and interpret the results in problem situations.
- Investigate limiting processes by examining infinite sequences and series and areas under curves.

And so that, in addition, students planning to attend college can:

- Understand the conceptual foundations of limit, the area under a curve, the rate of change and the slope of a tangent line, and their applications in other disciplines.
- Analyze the graphs of polynomial, rational, radical, and transcendental functions.

Standard 14: Mathematical Structure

In grades nine through twelve, the mathematics curriculum should include the study of mathematical structure so that all students can:

- Compare and contrast the real number system and its various subsystems with regard to their structural characteristics.
- Understand the logic of algebraic procedures.
- Appreciate that seemingly different mathematical systems may be essentially the same.

And so that, in addition, students planning to attend college can:

- Develop the complex number system and demonstrate facility with its operations.
- Prove elementary theorems within various mathematical structures, such as groups and fields.
- Develop an understanding of the nature and purpose of axiomatic systems.

Content of the High School Program

Strands in Grades Nine Through Twelve

The strands have already been described in Chapter 3, pages 80–87. However, the scope and complexity of high school mathematics demand that the strands be described here in greater detail. These descriptions specify the scope of the mathematics content students are expected to study in the core curriculum and in the post-core curriculum—either in subsequent courses or as additional work in the core curriculum courses themselves.

In the description of strands, certain topics appear in more than one location to emphasize the fact that certain ideas will reappear several times with different emphases and levels of abstraction. The Pythagorean theorem will appear in both a geometric and an algebraic context, and polynomials will be discussed at both an elementary and a more advanced level.

Functions

Functions allow the representation of relationships among quantities simply and efficiently. The concept of function is so important that the majority of high school mathematics can be seen as built around functions: developing the tools of algebra (to define and utilize functions) and of graphs (to visualize the important characteristics of functions); studying important types of functions, such as linear, exponential, and trigonometric functions and their typical applications; and applying functions in areas such as geometry, discrete mathematics, statistics, and probability.

Functions in the core curriculum. The essence of the study of functions in high school involves characterizing the *relationship* between two quantities (or other entities) and developing a useful sense of the ways in which one quantity can *vary* in relation to another. This study proceeds through the building up of the basic notion of a function, the relationship and meaning of input and output for functions, and the comparison and translation among various ways of describing functions: verbally, graphically, algebraically, and numerically in tables.

An important role of functions appears in the modeling of phenomena, such as growth and decay, with emphasis on the variety of phenomena that can be represented by the same type of function; and in the analysis of characteristics, such as direct and inverse variation in terms of the modeling situation. The work needs to be made specific and concrete through discussion of particular elementary examples of functions: linear, polynomial, exponential, quadratic, and simple rational.

A fuller understanding of the function concept comes through introducing the notion of the inverse of a function, illustrated graphically in tables and with a calculator, and through bringing out the role of function inverses in solving equations. Determining and visualizing graphical representations of functions are also crucial. Both quantitative and qualitative analysis of the graphs of functions are important and involve the notion of slope, x and y intercepts, and the general shape and position of the graph. Further development of the function concept and notation proceeds through introducing the ideas of domain and range, the algebra of function composition, and piecewise-defined functions.

An important aspect of understanding classic families of functions is how they connect to the behavior they typically model. Why are trigonometric functions useful for modeling periodic phenomena? Why are exponential functions useful for growth and decay situations? The behavior of different kinds of functions and their graphs, especially asymptotic behavior, minima and maxima locations, and intersections with the axes, becomes part of the students' *variation sense* or *relationship sense*. This development is especially important for exponential, logarithmic, and trigonometric functions.

All of this work with functions needs to proceed together with an emphasis on graphical representation, calculator evaluation and applications, and the ability to form visual images of graphs of common functions defined analytically or algebraically.

Functions in the post-core curriculum. The post-core curriculum includes the introduction of linear transformations in the plane: rotations, reflections, and similarity transformations; their representation by matrices and their geometric properties; and certain nonlinear transformations, such as translations. There is the development of graphical representation of parametrically defined functions with the aid of a graphing calculator, including functions defined in polar coordinates, graphical representation of translation in the range and domain of functions, and interpretation of parameter changes. Finally, there can be an intuitive introduction to the notion of a limit of a function, using as an example average and instantaneous velocity and calculator computation of successive approximations to limits, including e and π.

Algebra

Algebra is a fundamental mathematical tool for thinking and communicating symbolically across all the strands. In their study of algebra, high school students need to develop *symbol sense*, a general fluency with using symbols to describe relationships among quantities and to make generalizations. Algebra makes the most sense in the context of applications in both mathematics and the real world, where its generalizing power makes situations easier to analyze and understand. Algebraic fluency should be developed in conjunction with work involving the use of functions, geometry, statistics, probability, and discrete mathematics.

Furthermore, high school students should develop a preliminary understanding of algebra as an expression of mathematical structure. Students can simplify and summarize their understanding of how the laws, rules, properties, and definitions related to arithmetic operations work together. The elegance of algebra can make the structure accessible to students who have had ample experience using algebra in concrete situations.

Algebra in the core curriculum. The notions of variables and symbol manipulation and the relationships between them are the core concepts of algebra. Student understanding and use of these concepts should continue in the core program through the description of quantitative situations by means of algebraic expressions, equations, and inequalities. Students should learn to interpret by means of tables, graphs, diagrams, and verbal descriptions. The ways in which standard forms of expressions, equations, and inequalities can be obtained (by symbolic transformations, using addition, subtraction, multiplication, and division; forming powers and roots; and factoring) should be exploited to increase the variety and breadth of interpretive methods students can use. Students should learn how the interrelationships between algebraic and geometric representations of equations and inequalities can assist in determining specific results in investigating problem situations. By using these techniques in conjunction with problem-solving activities, students will learn to appreciate the power of algebra as a concise language for describing and modeling problem situations and as a tool for solving specific problems.

The further exploration of polynomials, with an emphasis on graphing and applications of the fundamental theorem of algebra, including the use of the connection between linear factors and zeros, will enhance the advanced students' ability to work with the concept of algebraic transformation and recognize classes of algebraic expressions. Working with these ideas will also enhance the students' ability to abstract in a mathematical context.

Algebra in the post-core curriculum. Students can experience the power of mathematics through the study of matrices as a class of algebraic objects that can be used to represent data and to represent information described in terms of linear equations. Student understanding of the power of algebra can be extended by the study of other mathematical subjects, including complex numbers, mathematical induction, the binomial theorem, and arithmetic and geometric series.

Geometry

Geometry needs to play a major role in each year of high school and should be developed in ways that are both broader and deeper than has been typical. The heart of high school geometry involves ideas about size, shape, location, and dimension that should be developed in both two and three dimensions and should be explored in the everyday world of the students. Thus, students need to see the analogy between familiar frames of reference used to locate places (street patterns, building interiors, and playing fields) and the study of coordinate geometry. Ideas about distance and direction should be developed both concretely and abstractly in synthetic and analytic contexts. Shape and size should be studied in the world, in computer environments, and in abstract contexts of geometric, numeric, and algebraic representation. Patterns in space need more attention. Symmetries, tessellations, fractal geometry, and packing problems are not only important but also fascinating for high school students. The basic concepts of topology are relevant as an extension of geometric ideas.

The goals of geometry in high school are to develop students' power to use geometric ideas and tools to understand and represent two- and three-dimensional situations. Included should be the development of students' reasoning skills in a geometric context, with emphasis on why certain geometric results are consequences of more basic geometric assumptions. A traditional presentation of geometry as solely a formal axiomatic system does not meet these goals.

Quite generally, being able to visualize and represent situations geometrically is a powerful thinking and learning tool in helping to understand concepts in the other strands and in helping to portray and analyze situations.

Geometry in the core curriculum. In high school, geometry continues the development of elementary descriptive geometry from earlier grades: the study of distance and direction through work with lines and angles, parallel and perpendicular lines, slopes, reflections and distance, angular measurement, and straight and vertical angles as well as the study of size through work with area, perimeter, and volume of various two- and three-dimensional figures.

Although the concept of congruence is important in high school, ideas revolving around similarity should receive even more attention. Understanding similar figures in general and similar triangles in particular is crucial in many applications within and outside mathematics. This work should lead naturally to the development of right-triangle trigonometry, with problems involving right triangles; similarity and invariance of ratios of sides; and sine, cosine, and tangent as functions of angles.

Students learn to use straightedge and compass constructions to solve problems and understand geometric relationships as well as to appreciate the impossibility of certain constructions. But students also explore a balanced repertoire of constructions with other instruments, including ruler and protractor, paper folding, mirrors, shadows, computer environments, and other physical manipulatives.

The Pythagorean theorem has many proofs that illuminate its truth in different ways. Students should use a variety of these proofs and should be able to construct a convincing demonstration for the correctness of the theorem. This theorem will be used in applications to determine distances and calculate angle measure.

Explorations of the circle, angles, arc length, conic sections, and tangents should be connected to algebra and trigonometry. The use of vectors and transformations (including translation, rotation, and reflection) can be integrated with functions and algebra to explore and solve problems in the plane and in space as well as in the applications to geometric constructions and arguments. Realistic situations and physical models, as well as the introduction to three-dimensional geometry and two-dimensional representations, will enhance the understanding of geometric models.

Certain nontraditional subjects, such as tessellations, patterns in space, crystal structure, packing problems, fractal geometry, computer graphics, and the basic ideas of topology, deserve a prominent place in the curriculum along with more traditional concepts, such as symmetries, projections, and sections.

Some principles of logical reasoning will be integrated within the geometry curriculum through the study of inference, hypotheses and conclusions, implications, and deductive arguments. However, students in the core should frequently approach proofs more from an informal standpoint that emphasizes their role in illuminating an idea than from a formal standpoint that emphasizes the logical structure of deductive systems. Therefore, proof of the obvious should be avoided in favor of proof as a way of seeing which nonobvious things are true. The format and technicality of proofs should be downplayed, and the verbalization and visualization of reasoning should be highlighted.

Geometry in the post-core curriculum. Beyond the core curriculum geometry should continue to be integrated with work in other strands; for example, with *algebra* and *number* through the geometry of complex numbers or with *functions* through the meaning of the slope of or area under graphs. Major emphasis should be placed on the development of coordinate and vector algebra and on their applications, frequently in the context of explorations and investigations.

Statistics and Probability

A working understanding of statistics and probability has become essential for active citizenship and for an ever-widening range of occupations. Using mathematics to make sense of uncertainties deepens students' *data sense* and broadens their appreciation of mathematics. Statistics and probability sharpen the students' power to reason from assumptions and evidence and evaluate confidence in conclusions. The mathematical ideas and methods of statistics and probability should be drawn from working with realistic situations, experiments, and games. Inferences from data (e.g., statistical quality control or opinion polling) in the social studies curriculum and in vocational programs merit special attention as subject matter for statistics and probability.

Statistics and probability in the core curriculum. Statistics and probability build on collecting, analyzing, representing, and interpreting data from real and simulated situations. Students understand and use the statistical and probabilistic points of view regarding uncertainty.

147

Many of the ideas and methods of statistics are based on the concept of frequency distributions. Students need a sense of how frequency distributions are created and interpreted in a variety of situations. Population, sample, and randomness are related key ideas. Students should use random sampling methods, understanding issues such as bias in sampling, sample size, and the reality that sampling can go wrong.

Using statistical methods, high school students should be able to design and carry out a simple but well-thought-out research project. Data can be collected by survey methods or experiment. Students should understand the practical meaning of experimental design, control group, quality control, replicability, and reliability. Techniques of data analysis should include measures of central tendency (mean, median, mode); dispersion (variance and standard deviation, range); and relationship (correlation, regression). An informal study and the use of inferential procedures, including hypothesis testing and confidence intervals, are also appropriate.

The construction, analysis, and interpretation of graphs, charts, and tables should be emphasized in the core. Bar graphs, stem and leaf plots, pie charts, histograms, frequency distributions, line plots, tables, matrices, and other methods of displaying and communicating data are essential tools. Misleading representations of data are critically important for students to recognize and avoid. In all this work, technology should play a prominent role. Software, including graphing applications, spreadsheets, data bases, and statistical packages, are basic tools for work in statistics. Students need experience with their use.

To make sense of experiments involving chance, students employ basic ideas of probability: randomness, independent events, mutually exclusive events, complementary events, equally likely events, and expected value. They can find the probability of an event, using ideas and methods that include relative frequency, simulation, sample space, tree diagrams, tables, and geometric models; and they can use the basic operations on probabilities, including conditional probabilities. By experiment and simulation the binomial distribution is constructed, and its application to real phenomena and random walks is developed. In addition, the difference between empirical and theoretical probabilities is clarified. Students employ counting principles and methods to solve problems that involve combinations, permutations, and Pascal's triangle.

In the core curriculum probability is developed in the context of its uses in interesting situations: odds, games, insurance, the lotteries, genetics, and forecasting. Because of the natural fascination these situations have for students, probability can and should be used as a means for recruiting many students for the community of people who enjoy and appreciate mathematics.

Statistics and probability in the post-core curriculum. Further development of statistical methods should include normal distribution and its relationship to the binomial, the central limit theorem, confidence intervals, statistical significance, and hypothesis testing. The meaning and use of correlation, regression, outliers, z-scores, standard deviation, and other summary statistics should be studied. Discrete and continuous probability distributions should be used to model real-world phenomena. Understanding of theoretical probability should be extended to include methods for determining expected value. Post-core work in probability and statistics can be especially worthwhile for students interested in humanities, social science, business, biological sciences, and technology.

Discrete Mathematics

The importance of discrete mathematics has increased significantly in recent decades, and the high school curriculum should reflect this fact. Discrete mathematics, the study of systems with separate (discrete) entities, is contrasted with systems involving continuous quantities. It is especially important in the information sciences and other situations in which the relationships among finite sets of elements are of interest.

Finite graphs and their associated matrix representations are important additions to students' repertoire of problem-solving tools. The diagrams, networks, and flowcharts that students construct to model situations or use for planning, scheduling, and decision making can be explored for their mathematical properties. Students need experience developing and analyzing algorithms in a variety of situations: practical (instructions, games, procedures); computer (programming, applications); and mathematical (prime factorization). In addition, they should investigate counting and arranging techniques and understand how these techniques are used in probability. Many of the ideas in discrete mathematics lend themselves to practical projects and integration with topics from other areas of the curriculum, especially the social sciences.

Discrete mathematics in the core curriculum. Realistic situations often have discrete features that can be represented by mathematical structures appropriate for high school students, particularly directed and undirected graphs and their associ-

ated matrices. Students can use these ideas to make sense of scheduling, routing, and networking; employ critical-path methods in practical contexts; and use simple methods to organize classifications as hierarchies or cross-classification. The idea of recursion is developed as a means of reducing complicated situations to simpler ones. Students construct, analyze, and compare algorithms and use graphical methods of linear programming and their discrete analogues as a decision-making technique. Combinatorial methods are used to analyze situations involving probabilities.

Discrete mathematics in the post-core curriculum. Students become familiar with elementary matrix operations and the implication of these operations in different situations. Mathematical induction is developed as a formal proof technique with a variety of applications.

Measurement

High school students develop more mature insights into the essential role of measurement as a link between the abstractions of mathematics and the concreteness of the real world. In high school most work with measurement connects conceptual developments in other strands to practical situations. Of central importance are the measurable properties of physical objects: length, area, volume, weight, and angle. Measures of time enable students to study the motion of objects. Units for these directly measurable properties (e.g., meters, seconds, and grams) can be combined to measure other properties (e.g., meters per second, grams per cubic centimeter, or cycles per second). The connections among compound units, formulas, functions, and the underlying physical properties should be a major component of students' mature understanding of measurement.

Measurement should be seen as one of the important answers to the question, *where do numbers come from?* Students should develop good judgment about using numbers that come from measurement, particularly a realistic sense of what is a meaningful level of precision. Students should also develop a critical eye for the validity of measurement; that is, how well does the instrument measure what it purports to measure?

Measurement in the core curriculum. In connection with other strands (and, where possible, with work in other subjects), students work with fundamental, measurable quantities of physical objects: length, area, volume, time, angle, and weight. They use a variety of instruments that measure these quantities and design and construct simple instruments of their own, using mathematical principles (e.g., trigonometric ideas for remote measurement). Work is typically done in metric units, and calculations and graphs are based on an increasingly sophisticated use of calculators and computers.

Students make use of measurement units from a wide variety of applications. They understand how units interact with calculations and with interpretation of results. In practical situations, often drawn from other subjects, students combine units to make new variables: (e.g., meters per second or person-hours). They understand how formulae and functions can give rise to such compound units and enable them to determine quantities that they cannot measure directly. Students make sense of converting quantities from one set of units to another. They understand and apply good judgment about the precision of measurements and know the effect of limited precision on the accuracy of calculations.

Measurement in the post-core curriculum. Beyond the core curriculum, students learn to use measurement in more complex situations that require more advanced functions and statistical models. They work with the idea of variation due to errors of measurement in the analysis of data. They explore with mathematics some of the more important and fascinating measurement problems from science (such as early attempts to measure the size of the earth and the distance to the sun and the moon, later efforts to measure the speed of light, and recent work on measuring the size of the universe); construction and architecture; the arts and crafts; and social science.

Number

Students' *number sense* must continue to grow throughout high school, and their idea of number must come to include such constructs as irrational numbers, complex numbers, and vectors. During the core years students should develop confidence connecting their number sense to algebraic and coordinate representations. They should also acquire a good feel for what operations and combinations of operations do to quantities. The basis for a deepening sense of number is experience with the variety of ways numbers are used to express size or amount and order (greater or lesser) and to identify objects in a collection. On this basis is built an understanding of how numbers are used in each of the strands.

Number in the core curriculum. Most work with number in high school is embedded in investigations and problems arising in other strands. Yet students extend their number sense from its grounding in quantitative features of the world to an appreciation of how number fits into a structure based on the arithmetic operations. Students develop insight regarding the multiplicative structure of integers (primes, factors, and multiples) and the multiplication operation on different kinds of numbers; deepen their intuition about the differing effect of the four basic operations, plus powers and roots, on numbers greater than or less than one and greater than or less than zero, thus highlighting the special roles of zero and one; clarify their view of the relationships among the operations; and, finally, work with the basic relationships among integers and rational, irrational, real, and complex numbers.

Students become familiar with different representations of numbers that arise in varied contexts: decimal, fraction, percent, per unit, integer, logarithmic, and scientific notation. A strong sense of what numbers mean in a particular situation is developed for quantities that have particular units (or "dimensions") and for dimensionless quantities such as ratios, coefficients, and scale factors. The order property of real numbers is given meaning through both practical and mathematical contexts.

Number in the post-core curriculum. After the core curriculum the development that students have experienced in other strands frequently returns the students to number with accompanying new issues. Complex numbers, algebraic and transcendental irrationals, vectors as two- and three-dimensional extensions of the concept of number, and work with specialized computational methods are all examples of how number sense continues to grow beyond the core.

Logic and Language

The primary goal of the logic and language strand at the high school level is to demonstrate how methods of logical reasoning framed in accurate but not excessively formal language can contribute to the clarification of ideas in all areas of mathematics as well as in other fields. The strategies needed to pursue this goal differ in several ways from what is common practice today.

Ideas of logical reasoning in high school have typically been too narrow—a result of being developed almost exclusively in the context of plane Euclidean geometry. The result is a situation in which students have only a cumbersome understanding of logical argumentation that is not useful to them in other work within or outside mathematics. Another casualty of this approach is a usable understanding of geometry itself. Insight and intuition about geometric principles seem to be sidetracked by the need to formulate these principles in a language inappropriately rigid and formal.

Logic should not be treated as a pure subject separate from any other strand. In particular, it is not generally suitable in the high school core to develop extended treatments of formal propositional logic (including truth tables of the connectives *and*, *or*, *not*, *if . . . then*, and *if and only if* and the role of these connectives in laws of inference). Instead, ideas of logical reasoning should be integrated with work in *all* strands in an appropriate and balanced fashion. Ideas from logic are appropriate when they are seen as tools that are helpful in gaining understanding rather than as rigid standards that represent a barrier to understanding. In particular, proofs are appropriate when they are developed as a way of explaining why certain assertions are true. They should serve to make clear things that were not clear before. However, they are often inappropriate when used as a painstaking, formal demonstration of obvious results.

Looked at another way, a proof is appropriate when a student produces it as a summary of an exploration and the proof clarifies the work; but it is inappropriate when a student memorizes it without being immersed in the particulars of what the proof is about. An appropriate tone can be set for the role of proof through illustrations of elegant and simple proofs that demonstrate and clarify nonobvious results.

Above all, logic in high school needs to be seen less as a discipline in which students work with formal logical systems and more as a discipline in which students extend and refine their understanding of everyday language in its capacity to express careful lines of reasoning. It is to emphasize this connection with language that the strand has been named *logic and language*.

Logic and language in the core curriculum. Throughout high school, students need work with *inductive reasoning*, interpreted broadly to include recognizing and extending patterns; classifying and sorting by attributes, similarities, and differences; reasoning through analogical comparisons; observing relationships and making generalizations; and enumerating and analyzing information by statistical means. In this work students need to understand the primary value of inductive reasoning (as a means of formulating hypotheses and conjectures) and

also its limitation (as leading to plausibility rather than to certainty).

There should also be many opportunities in high school for students to see and use *deductive reasoning* in ways that include understanding the role of hypotheses and conclusions; appreciating the difference between a statement and its converse, its inverse, or its contrapositive; being clear about the need to state and validate assumptions; making use of explicit definitions in the context of a proof; recognizing flaws in deductive arguments; using direct proofs and understanding the logic behind the typical step ("Let X be a . . . ") in a theorem that makes claims about X; and using indirect proofs and understanding the logic of counterexamples and contradictions.

It should not be the goal of the core program to develop students' proficiency in constructing or memorizing proofs. Rather, students should develop their reasoning power so that they can recognize the breakdown of a logical argument due to valid reasoning from faulty assumptions or to faulty reasoning from valid assumptions and appreciate the fact that small differences of assumption can lead to large differences in conclusion.

This work with inductive and deductive reasoning needs to permeate work in all strands and be carried out by having students frequently discuss their reasoning about the mathematical subject matter, orally and in writing, for themselves and for an audience. This work needs to be supported by active study of relevant and useful logical features of language itself. Such study should lead to familiarity with such common logical concepts as necessary conditions, sufficient conditions, conditional statements, implicit assumptions, inconsistent premises, and the like; with the special meanings of such connective words as *not, and, or, if . . . then . . . , whenever . . . then . . . , . . . only when . . .* , and *. . . implies . . .* ; and with the various (and sometimes implicit) uses of "quantifying" words such as *some, none, any, every, all,* and *each.*

Finally, the study of logic in high school needs to extend to discussion of the special uses in mathematics of many words and of the relationship to their meaning in everyday language. Some of many examples are *rate, variable, function, scale, slope, dimension, formal, abstract, general, ratio, proportional, proportions, in proportion, quantity, fraction, rational, unit, parameter, root, model, expression, fixed, necessarily, alternate, corresponding, complement, extremes, origin, magnitude, reciprocal,* and *equivalent.*

154

Logic and language in the post-core curriculum. Appropriate for study here are the more technical aspects of logical reasoning, such as mathematical induction; the use in mathematics of formal deductive proof in axiomatic systems, illustrated with a review of the role played by the development of such systems in the history of mathematics; and the effect on other disciplines of the success of some axiomatic approaches as well as their inherent limitations.

Connections Across Strands

Questions related to program content and course structure require an explicit and forceful statement as to the connections across the eight strands. This statement is provided in two ways. Previously, a description of each strand was presented to state the essence of each strand and its major connections to other strands. Now certain connections across categories that are so important as to require special emphasis are discussed separately:

Connection between functions and algebra. In high school, algebra is too frequently viewed as a subject distinct from functions. Many current introductions to algebra do not mention functions and their graphs at all or do so only in an unconnected way, often toward the end of the course. As a result, when they study functions and their graphs, students lack a visual and conceptual backbone that can help them to make far more sense of algebra and to find ways to put algebra to immediate use. Further, functions themselves are often seen as *hard* and *advanced* subject matter in high school. They are presented in unnecessarily abstract and formal terms and hence are not learned by students as useful and usable ideas.

It is far more worthwhile to view algebra as a symbolic language whose primary purpose is to allow students to understand and use functions. Specifically, the language of algebra is needed to construct and manipulate the expressions that define functions, and the techniques of algebra are needed to solve the equations and inequalities that result from stating relationships between these expressions. In this conception algebra is the symbolic language that underlies work not only with algebraically defined functions such as polynomials and rational functions but also with transcendental functions such as logarithmic, exponential, and trigonometric functions; with distributions in probability and statistics; with recursively defined functions of discrete mathematics; and with the algebra of geometric transformations.

Instructional materials will be lacking if the intimate connection between algebra and functions is not made clear and unavoidable.

Connection between functions and geometry. The connection between functions and geometry is also strong. After all, graphs are a geometric vehicle for representing functions; and ideas such as slope, area, distance, and scaling that are important in understanding graphs of functions are concepts developed and used in geometry. Conversely, the important quantitative relationships of geometry involving lengths, areas, and volumes can be expressed clearly in the language of functions through the functional relationships of the relevant parameters. For example, the volume of a right circular cylinder is a function both of the height and of the radius of the cylinder; it varies directly as the height and directly as the *square* of the radius.

Another merging of the strands of functions and geometry comes through geometric transformations, such as reflections, rotations, expansions, contractions, and translations. These transformations are special kinds of functions that operate so as to move or change geometric figures. They bring order and insight to the study of properties of two-dimensional and three dimensional figures.

Connection between algebra and geometry. Standard number 8 of the NCTM *Standards* is "Geometry from an Algebraic Perspective." It encompasses subjects often gathered under the headings "Analytic Geometry," "Coordinate Geometry," and "Transformation Geometry." The subjects in this category represent an extremely fruitful and important merging of the strands of algebra and geometry. It is crucial that these subjects be developed along with the ideas of algebra and geometry from the beginning rather than be postponed until later courses. Emphasizing this link has an additional advantage; it makes less likely the situation that algebra and geometry will continue to be treated in isolation from each other.

The connection between *algebra* and *geometry* goes much deeper than coordinate geometry. An interesting way to observe this connection is to examine the variety of demonstrations titled "Behold!" or "Proof Without Words" that have been appearing in journals in recent years. These demonstrations frequently give a geometric interpretation that clarifies an algebraic statement and makes it more obvious. This kind of use of geometry in algebraic work needs to be seen more often in high school.

Conversely, it is frequently useful in geometry to set up relationships algebraically, as in dealing with relationships among several angles in a geometric figure. In fact, an extremely common situation that occurs when mathematics is being used involves a kind of union of algebra and geometry with functions. Thus, a diagram is set up to depict the essentials of a problem; geometric principles are used to figure out the distance and angle relationships in the diagram; and algebraic principles are used to set up the functional relationships that lead to the desired analysis of the problem.

Connection of trigonometry to algebra, geometry, and functions. The strong connections involving algebra, geometry, and functions have already been discussed in a general way. A more specific arena in which these three strands merge is trigonometry. Although the ways in which these four subjects are related may seem obvious, many aspects of their relationship are never utilized in school mathematics. Consequently, students often see the four as merely different subjects they have to learn. The connections among the four subjects need to be brought out continually in the work students do.

One example of the intricacy of the relationship is presented here. A little thought about the SAS postulate for triangle congruence of synthetic geometry reveals that this postulate has a trigonometric counterpart in the Law of Cosines:

$$a^2 + b^2 = c^2 + 2ab \cos\theta.$$

This law itself can be seen as a generalization of the Pythagorean theorem to deal with all triangles, not just right triangles. With algebra the Law of Cosines can be reexpressed as the function

$$f(a,b,\theta) = (a^2 + b^2 - 2ab \cos\theta)^{1/2},$$

which gives a prescription for finding the missing opposite side c of a triangle, given SAS for that triangle (i.e., given the side-angle-side quantities a-θ-b), and shows specifically how SAS determines a unique triangle.

Ideas from trigonometry need to grow continually along with the development of algebra, geometry, and functions. Students can use the language of algebra early in geometry when they express the ratios of side lengths that arise in dealing quantitatively with triangles and when they show how these ratios are functions of particular angles. In later explorations of circular motion, these same functions appear again as the sines and cosines that continue to be of central importance in the mathematics of periodic phenomena and complex numbers.

Other connections. The kinds of connections between strands discussed previously are crucial to the development of a coherent program in high school mathematics. But this listing of connections is not meant to be complete because many other connections are also important. The connection between discrete mathematics and probability has often been pointed out. The kind of counting arguments developed in discrete mathematics is the direct basis for forming the ratios that characterize probabilities in many situations. A connection between geometry and probability appears in the study of situations involving an object landing randomly inside or outside certain geometric regions. Such situations lead to a general geometric model for probability that is helpful in visualizing even nongeometric situations. The connection between probability and statistics is shown through their many shared features, such as the concept of randomness; the role of large numbers (of trials or elements of a data set); and the use of certain distributions, such as the normal distribution.

As a final example, it is worthwhile noticing how probability and statistics rely heavily on the tools of algebra and functions. This case and similar other cases represent, in a sense, applications of mathematics within mathematics itself.

Unifying Ideas for High School

The mathematical content for high school must be organized to develop depth as well as breadth. For depth to be achieved, a small number of unifying ideas must be chosen as priorities for the core program. A healthy discussion about which big ideas should unify the core program in high school is just beginning. This *Framework* endorses the debate as important for teachers as well as program designers.

In the past too many books and programs failed to pull the pieces together and take the time to develop important ideas deeply. As a result, many students experienced mathematics as miscellaneous details without underlying unity or connection. Therefore, core programs and the material that support them should be organized to develop depth for three to six unifying ideas. Suggested ideas are presented as follows:

Mathematical modeling. Mathematical modeling enables students to appreciate the power of mathematics to help them understand the world. Because most real situations are too complex to be modeled exactly, the limitations of modeling are gradually appreciated by students. Understanding those limitations is one of the most important things to learn.

Mathematical modeling moves first from a real-world situation to the realm of mathematics, extracting key features of the situation and translating them into mathematical terms. After relevant variables are identified, the relationships among those variables are described in suitable mathematical terms. Next, a mathematical analysis for the purpose of gaining insight into the nature of the phenomenon is developed and explored. After this mathematical formulation of the problem is worked on, a move is made away from the model and back to the real-world situation to see what insights have become possible through the power and precision of mathematics.

Sometimes this modeling is straightforward; that is, the mathematics arises directly from the situation itself. For example, vectors are used to model a boat moving in a current; exponential functions, to model population growth; and trigonometric functions, to model sound waves. At other times statistical techniques, such as regression, are used as an aid in understanding data from real phenomena. This modeling frequently requires that the data be transformed by the application of an appropriate function to the data to make a relationship appear linear. For example, logarithms of world population data might be fitted against time to make a better prediction.

Processes such as gambling, insurance, and genetics can also be modeled through probabilistic simulation. In any case mathematical models are used for a purpose; that is, to understand the past or predict the future. Inconsistent or unrealistic results may compel a return to the real-world situation to reformulate the model.

Variation. How is a *change* in one thing associated with a *change* in something else? A formula, such as the one for the surface area of a sphere, $A(r) = 4\pi r^2$, precisely captures a mathematical relationship but does not in itself give a way of thinking about what this relationship *means*. Thinking in terms of *variation* can help students use the idea of related changes to understand mathematical relationships. An increase in one quantity results in an increase (or decrease) in another. Students look at such variation empirically and analytically in a wide variety of situations.

Many important situations require more than two variables. For example, in compound interest the formula $F = E(1 + i)^n$ expresses a relationship among four quantities: the fortune F of money—originally a nest egg, E—invested for n compounding periods at an interest rate i per period. The formula is used to develop the idea of variation so that the situation can be understood better. For example, if the amount of money (F) wanted and the amount (E) possessed are known, it can be asked how the number of months (n) one needs to wait depends on (that is, varies with) the interest rate i. Algebra is used to reexpress the formula as $n = \log(F/E) / \log(1 + i)$. What does that expression mean? As the interest rate increases, the number of periods goes down. A more sophisticated analysis shows that, for small interest rates, doubling the interest rate halves the savings time.

Technology can play a role in exploring variation to clarify both numerical and geometric relationships. For example, students may fix some features in a bridge truss design and study how a change in a particular angle between truss members affects the truss as a whole. Or they might explore intersections of the altitudes of a triangle as the parts of the triangle change. These are complicated relationships; interactive geometry and design software can display the variation. Numerically, spreadsheet and graphing programs can calculate and display relationships like those in the compound interest example above. Some common formulas cannot be solved for an important variable; spreadsheets can help with numerical solutions when analytic ones are unavailable.

This way of looking at functional relationships extends across many strands and is not limited to situations that can be expressed in formulas. There are situations in which variations in a quantity for different categories are to be compared. For example, a survey of student heights might be made. What is the relationship between the distribution of those data and, for example, grade level? How does variation within grades compare with variation between grades? The idea of variation leads to a deeper understanding of both the distribution and the idea of standard deviation.

Sometimes the important result is a *lack* of variation. If a triangle is changed by keeping the base fixed and moving the opposite vertex parallel to the base, the area does not change. This is a case of *invariance*. As a final example the relationship between a matrix and its determinant can be considered. Some changes (multiplication of a row by a constant and the adding of the result to the next row) are associated with no change; other changes (interchange of columns) involve only a change of sign; and still other changes (duplication of a row or column) result in a zero determinant.

Understanding variation requires important skills and helps students use them in meaningful contexts. For example, the need to understand variation motivates learning when and how to solve for one algebraic quantity in terms of another. Controlling variables is essential for dealing with real data. The question of *optimization* appears when one entity is varied until an optimal value of the other entity arises.

Students can develop a unified view of mathematical relationships of many different types if they consider the relationships in terms of variation, asking how particular changes in one entity are associated with the corresponding changes in related entities.

Algorithmic thinking. The traditional school mathematics curriculum has justifiably been criticized for overemphasis on the memorization and rote application of algorithms. If the focus is placed instead on the process of *developing, interpreting,* and *analyzing* algorithms, students can develop insight into the nature of mathematics. Some considerations worth noting here are the following:

1. *Development of algorithms.* Middle school students learn to identify and describe patterns. In high school this kind of activity can be extended to include description and analysis of the steps and procedures that lead to the solution of many kinds of problems. In this way students can begin to develop their own algorithms.

2. *Analysis of algorithms.* Students may develop more than one algorithm for the solution of a problem or a particular class of problems; therefore, it makes sense to compare the algorithms. Students can experiment with the implementation of these algorithms, analyze their effectiveness, assess their efficiency, and discuss their relative appropriateness in specific cases.

3. *Role of technology.* Technological advances continue to change the kinds of problems that are important and the techniques by which the problems are solved. Although some aspects of the traditional algorithms of school mathematics are becoming less important, others are becoming more important, particularly the role of algorithms and procedures in using computer technology.

4. *Algorithms, understanding, and communication.* Algorithms are often used to understand important mathematics and to communicate this understanding to others. They pro-

vide a systematic structure in which the fundamental elements of mathematics are organized and their interconnections displayed. The ability to develop and interpret algorithms is, therefore, an essential ingredient in modern mathematical communication.

Mathematical argumentation. Mathematical argumentation helps students answer questions; for example: "How do I know when I'm right?" "Is this statement *always* true?" "Is it possible to satisfy these conditions?" In this way individuals are helped to make sense of mathematics for themselves and for others. A mathematical justification helps reveal the implications of the givens of the problem, the way in which things fit together, in a natural way. A formal proof, in an appropriate context, can illustrate the power of the axiomatic method in mathematics. However, this unifying idea extends beyond specific justifications and proofs to general consideration of what the standards and conventions of argumentation in mathematics are and how they differ in different areas of mathematics. Knowing the power and limits of statistical inference, for example, and expressing the inferences correctly are just as important as analytical proof. Relevant matters to be considered here are as follows:

1. *Social context.* Mathematical justification has a social context as well. It communicates one's understanding and helps others make mathematical sense of the world. It is a process similar to writing whereby author and audience share a common vocabulary and a similar frame of reference. A single basic idea might be expressed in many different ways according to the needs and experience of the audience. Mathematical discourse provides a channel for precision of expression and clear communication, including discussions of what is or is not accepted by members of the group.

2. *Movement from examples to mathematical arguments.* Empirical observations help students uncover relationships but are not in themselves sufficient to determine whether or not something should always be true. Working with examples is an important part of mathematics, but examples can only suggest. Mathematical arguments—from reasoned justification to airtight proof—convince us and others of the truth of our conjectures.

3. *Public process.* Mathematical argumentation should play a central public role in every class. The act of symbolizing

and arguing mathematically should be part of the process of sense-making and not simply an after-the-fact confirmation. It should be a public process to which all students contribute so that all students have reason to believe the result.

Mathematical argumentation provides a fundamental means of coming to grips with a mathematical object or phenomenon—a process that lies at the core of mathematical sense-making. This kind of sense-making is hard, but the power it provides is the ultimate justification for learning mathematics. When mathematics is learned correctly, mathematical argument is universal because it is necessary to achieve both individual and shared understanding.

Multiple representations. Using several kinds of representations provides a powerful tool for investigating mathematical ideas. Students are helped to clarify questions and situations, evaluate solutions, make mathematical connections, and elucidate underlying structure. Visual representations are especially powerful because they help make abstract ideas concrete and help students communicate with one another. Electronic forms of representations (e.g., simulations, graphs, symbols, and charts on graphing calculators or computers) facilitate exploration and transformation.

It is not enough, however, for a student simply to see and produce multiple representations. Real understanding comes when the student learns how to use and choose representations to clarify and communicate. For example, a student who has learned about graphing quadratic functions shows mathematical power when he or she moves fluently between the graph and the equation to help find solutions to a quadratic equation or inequality and understands the implications of these solutions. Similarly, a student confronted with a probability problem may use a tree diagram to enumerate the possible outcomes, but an area model to help calculate the probabilities. Students may differ in the representations they choose or find most useful.

An appreciation of multiple ways of representing an idea as well as a recognition of the equivalence of different representations helps lead students to deeper understandings of mathematical structure and process. Experience using different representations for the same situation also leads to a useful intuitive understanding of the idea of *isomorphism*, which is so important in mathematics.

Sample High School Unit from a Core Program

This sample unit has been used as part of the second-year core course in a successful integrated mathematics core program in California.[1] In this way the reader should have some idea of the scope and richness possible in high school units. In this core program each unit is developed in relation to a central question. The program is designed to teach some trigonometric functions in Course 1 (as ratios in similar right triangles) and to introduce the Pythagorean theorem in Course 2. Other sequences might avoid trigonometry until Course 3 but might introduce the Pythagorean theorem earlier. The syllabi for the core courses are explicitly not set so that innovation and diversity in curriculum development can be encouraged. Student work follows.

Overview of "Do Bees Build It Best?"

In this unit students work on the following problem:

Bees store their honey in honeycombs, which consist of cells they make out of wax. What is the best design for a honeycomb?

Students learn about the meaning of *area,* including the idea of a *unit.* They discover some basic formulas for calculating the area of specific shapes. Then they learn about the Pythagorean theorem. Using that theorem and the trigonometry that they learned in the first year's curriculum, students find a formula for the area of a regular polygon with a fixed perimeter. The formula is written in terms of the number of sides. Using the graphing facility on their calculators, students discover that the larger the number of sides, the larger the area. The circle is a limiting figure for the sequence of regular polygons having a fixed perimeter.

Students turn their attention to volume, using an approach similar to that used with area. They focus on the volume of prisms with regular polygons as bases. Then they find that the question *Which prism has the largest volume for a fixed lateral surface area?* reduces to the question *Which prism has a base with the largest area for a given perimeter?*

Using their earlier findings on regular polygons, students see that cylinders might be a candidate for the best bee cell. This candidate is rejected because cylinders leave gaps when they are fitted together. Students then look at the question *Which prism that has a regular polygon as a base and tessellates (i.e., leaves no gaps) will have the largest volume for a fixed perimeter?*

They find that cells in the shape of regular hexagonal prisms, the choice of the bees, are the mathematical winner if they consider only prisms with regular polygons as bases. Students conclude the unit by watching the video titled *The Mathematics of the Honeycomb* and discussing other candidates for cells besides prisms and other bases besides congruent regular polygons.

[1]The unit titled "Do Bees Build It Best?" was developed as part of the Interactive Mathematics Project, Lawrence Hall of Science, University of California, Berkeley, in cooperation with San Francisco State University.

Sample Assessment for "Do Bees Build It Best?"

The following questions were given as a take-home assignment; they are one of a variety of assessments for "Do Bees Build It Best?" A student response to this set of questions appears on pages 166–67. Another student's portfolio cover letter appears on page 168.

The city of Euclid is building a public swimming pool. The architect of the pool is a geometer at heart and decides that the pool should be built in the shape of a prism whose base is a regular heptagon (seven-sided polygon).

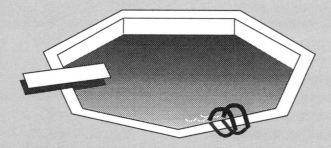

The pool will be four feet deep everywhere and will have a perimeter of 350 feet.

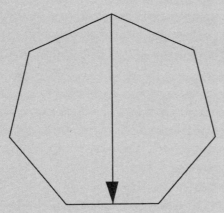

a. After the pool is built, the city will have to paint the inside of the pool—the bottom and the seven walls. Find the total surface area of the inside of the pool.

b. Once the pool is painted, it will be filled with water. Find the volume of water needed to fill the pool.

c. If a swimmer swims from one corner of the pool to the center of the opposite side as shown in the diagram, how far will that distance be?

Take-Home Do Bees Build it Best? assessment

Part I. Heptagonal Swimming

a. First I need to find the lateral surface area.

$350 \div 7 = 50$ — Lateral Surface
$50 \times 4 = 200$ area = 1400 sq. ft
$200 \times 7 = 1400$

Then I need to find the area of the base.

$360 \div 7 = 51.428571$

50

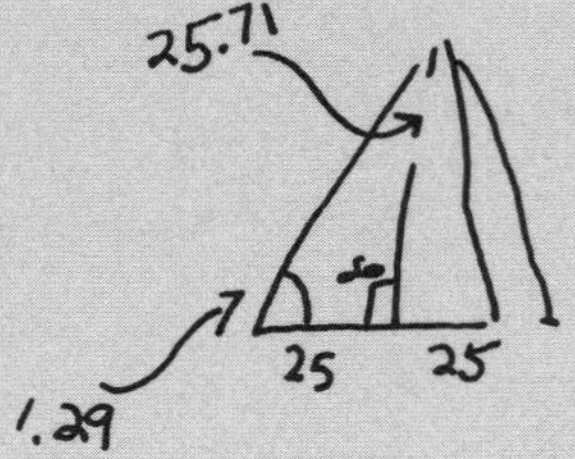

$Tan = \frac{opp}{Adj}$

25.71

25   25

1.29

$Tan \; 64.29 = \frac{h}{25}$

$2.077 = \frac{h}{25}$

$h = 2.077 \cdot 25$

$h = 51.925$

Heptagon Surface Area

$51.925 \cdot 50 \cdot \frac{1}{2} = 1298.125 \cdot 7 = 9086.875$

then I add the 2 surface areas together

$9086.875 + 1400 = 10486.875 \; sq. \; ft$

b. Since I already have the area
of the base (area of the Heptagon), all
I need to do is multiply it by the
height.

$$9086.875 \cdot 4 = 36347.5 \text{ cubic feet}$$

c. For this problem I will have
to cut the heptagon into triangles. The
distance will be one side of the
triangle plus the height of a triangle

I have the height. I need
to find the length of the
side

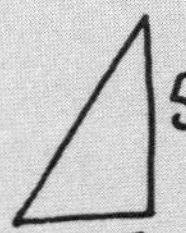

$$A^2 + B^2 = C^2$$

$$25^2 + 51.925^2 = C^2$$
$$625 + 2696.20 = C^2$$
$$\sqrt{3321.2} = C$$
$$C = 57.63$$

Now all I have to do is add
them together:

$$57.63 + 51.925 = 109.555 \text{ ft}$$

Dear Reader,

this unit was interesting, because we related all our work to bees. Yes, honey bees, We learned about tesselation and ended up relating that to M.C. Escher It was fun. because we got to do our own tesselations; it was a neat way of approaching geometry. Back to the bees — so we were faced with the question each day: "Do Bee's build it best?" referring to their hexagon shaped honeycombs. So we went through this whole thing of trying to tessellate circles, pentagons, octagons, dodecagons etc. meanwhile we were learning how to find the area & surface area and volume of these shapes. We worked with blocks, and actual honeycombs. This unit was un fun, I love reading the names and situations on each homework, for example: "Keisha is making a patchwork quilt" and "Rancher Gonzales is building a corral" & "Rasheed's dog is in a ditch." It's quite refreshing.

                    Jaeinda

168

Suggested Units for the High School Grades

In this section some ways of organizing mathematics into coherent units for high school are suggested. These suggestions are neither exclusive or mandatory; they are only possibilities for consideration. Some of them could be revisited several times during the core sequence. The suggestions are centered on *mathematical subject matter;* a unit might be *about* this mathematics but could never cover it. Most units will also have rich contexts not specified here.

How are the curricula of the past connected with what this *Framework* proposes? One of the appropriate questions to be is asked is, "Where, in this new curriculum, will students learn what they learned in algebra?" (The same question might be asked for geometry.) Part of the answer can be found in units that focus on the topic "Representing and Analyzing Physical Change over Time," which appears in the next part of this section. These units should provide a context and rationale for expressing situations symbolically and graphically and learning to manipulate expressions. For example, in such a unit students may well need instruction they will receive in algebraic mechanics. Just what instruction they will receive will depend on what they have seen before. Regardless of the level at which instruction in algebraic mechanics is offered, students will serve themselves well by learning it.

The unit just described should not be the *only* unit considered. The next suggested topic, "Measuring Inaccessible Distances," can serve a similar pedagogical purpose. Other units will have to be created as well so that students—over a period of time and in diverse, meaningful contexts—develop facility in using algebraic tools and techniques. That facility should be sufficient for students to achieve a depth of understanding of the unifying ideas.

Representing and analyzing physical change over time. This subject matter focuses on analyzing and understanding ways of graphing and describing motion or physical change over time. Students learn to sketch and interpret graphs of change taken from real-world phenomena and to generate ideas for real phenomena that might match a given graph.

Types of functions include constant rate (linear growth), constant rate with stopping, accelerated motion, circular motion, spiral growth, periodic motion, and conic sections from the locus point of view—generated from situations such as cars moving, balls bouncing, planets orbiting, and plants growing.

Developers evaluating these units as possibilities and devising their own should be sure to use the criteria for evaluating units that appear on pages 91–92.

Students become fluent in shifting between situations and graphs. A graphical representation of motion and a diagram showing the path of the motion itself should not be confused. For example, a ball thrown vertically traces a parabola only on a graph, with time as one axis.

The more experienced the students are, the better they will be able to use symbolic representations of functions: Cartesian, parametric, and polar. This subject matter also provides an opportunity to develop facility with the algebra of functions in meaningful contexts. And as students develop their facility with both graphical and symbolic representations, they become better able to create their own mathematical models for physical change.

Measuring inaccessible distances. Students use and develop mathematics to measure distances *to* and lengths *at* other places without going there. Problems can range from measuring the height of someone standing across the room to estimating distances from a moving vehicle to measuring the radius of the Earth.

As students mature, they can bring more and more mathematics to bear on these investigations, beginning with similar triangles and proceeding through trigonometry. This material also requires students to measure angles, linear distance, and time (for moving-parallax situations); use standard and nonstandard units; and deal with varying degrees of precision. Students produce solutions and reports ranging from simple answers (with estimates of possible error) to clear explanations of measurement techniques to new rules of thumb and their mathematical justifications. They develop and communicate their understanding through scale drawings, sketches, mathematical formulas, and graphs of functions.

Students create mathematical models for distance-finding. They explain these models coherently and write useful descriptions of how to use procedures they have invented.

Units about measuring inaccessible distances are also natural places to develop the unifying idea of *variation*. In the viewing of a distant object, for example, three quantities depend on one another: the distance to the object, its size, and the angle it subtends.

Families of functions. Studying this more abstract subject matter, students look at interesting properties of families of functions and their members. They reinforce their mathematical understanding of the functions they have seen in more concrete units and connect situations with one another through family resemblances.

We assume throughout that students have access to graphing utilities. Today, that generally means access to graphing calculators. With these utilities the youngest students in the program can study and use functions that were previously restricted to the oldest. And all students can make more sophisticated explorations of functions and their relationships.

Each core course can revisit families of functions. At first, students might take a qualitative look at

$$y = f(x) \text{ and } y = A f(x) + B \text{ for } y = x, y = x^2, y = x^3,$$
$$y = 1/x, y = \sin(x), \text{ and } y = 2^x.$$

They describe the shapes of the functions; see where they increase and decrease; study intersections and inequalities (over what interval is $x^2 > 2^x$?); investigate and compare slopes; and explore the effect of the constants A and B.

More experienced students can explore sums and differences of functions, horizontal translations, rational functions, and additional families, such as $y = \tan(x)$, $y = \sqrt{x}$, and $y = |x|$. The most experienced high school students can explore asymptotic behavior, learn to sketch complicated functions *without* the calculator, build functions from other functions, and explain why one function might dominate another.

In the context of exploring graphs and recalling the situations (from other units, generally) in which functions have arisen, students also develop facility in manipulating and understanding their symbolic representation. For example, younger students solve linear equations for x and graph the inverses; advanced students show both graphically and on the unit circle that $\sin(x + \pi/2) = \cos(x)$.

These units and their attention to abstract connections must build on *previous* experiences with functions and refer to them frequently.

Mathematics of geometrical solids. Students explore solid figures: the five Platonic solids, prisms, other common polyhedra, cones, and spheres. They find the solid figures in everyday life, construct them, and investigate their mathematical properties. These activities are the natural high school extension of middle grade units about *objects, shapes, and containers* (see pages 125–26). As in middle school, students use different representations (ranging from physical models they build

themselves to computer graphics) as they study these figures; and their attention is focused both on the object and its interior. Students continue to develop their capacities to visualize, represent, and communicate important mathematical properties, such as surface area and volume, plane and dihedral angles, parallelism, and perpendicularity.

But their understanding goes beyond what we expect of middle school students. They develop a more sophisticated understanding of the functions that appear. For example, they can create an expression for the volume (and surface area) of a truncated cone and can describe why they might not be able to do the same for a sphere. More experienced students can create an expression that describes the height of fluid draining at a constant rate from a pyramidal tank as a function of time and can sketch the graph for a tank of any shape.

They compare surface-to-volume ratios of different solids and develop and use mathematics of scaling to study that ratio as linear size changes. Rational exponents occur naturally in this area of study.

Some classical topics from solid geometry appear here as well, ranging from classification to proof. For example, students can describe why a rhombic dodecahedron is not a Platonic solid; can decompose a cube into pyramids; and can argue convincingly that there are only five regular polyhedra.

Looking back at triangles. In the review of triangles, experienced students summarize and reflect on the role of triangles in their mathematical development and in various types of problems. Specifically, students learn to tell whether they need to use the Pythagorean theorem, similarity, or trigonometry to solve a triangle in a particular situation *and to explain why*. They can also apply and explain triangle inequality and the sum-of-the-angles principle.

Students generalize their prior experience and see that six quantities specify a triangle completely on the plane; that there are different ways—for example, coordinates of vertices, or sides and angles—of choosing those quantities; and that there may be strong constraints on possible combinations. They explore which of the six quantities are sufficient to ensure similarity or congruence between triangles and allow deeper and reinforcing discussions of similarity and congruence. One such discussion leads to the realization that similarity appears much more frequently than congruence in real-life problems.

Another discussion exposes the connections between SAS and the law of cosines and between SAA and the law of sines.

Students become fluent in choosing to represent triangles and identify their components in a way most appropriate for the problem at hand. Ultimately, they develop problems of their own and readily recall situations in which a given way of dealing with triangles is most salient. They also identify situations (e.g., on a sphere) in which the plane rules do not hold.

Secrets and information: encryption and decryption. Students create codes and ciphers and develop techniques for transforming clear text into code and back again. They use many different ways of representing their ciphers, ranging from code wheels to matrices. And they learn and invent many different ways of encoding text, from simple number-substitution ciphers to book codes and beyond.

Mathematically, the study of discrete functions and their inverses and the development and analysis of algorithms are involved. Students find out what happens when a function's inverse is not a function: decryption is not unique. They also learn about composition of functions, (e.g., when they encode a coded message). In describing how to decode singly and multiply encoded messages, they develop understandings that underlie mathematical operators, transformations, modular arithmetic, and group theory.

Students use statistics when they try to decode messages without knowing how the messages were encrypted. They analyze text in various languages to see the distributions of words, characters, and character-pairs to aid in the decryption and creation of more efficient encryption and compression algorithms.

Manual and computer-based simulations help students explore the effect of imperfect transmissions. Students come to see codes not just as a means of sending secret messages but as information preparing them for the inevitable introduction of information theory into the fundamental college curriculum.

The items discussed in this section are only examples. Other chunks of subject matter that have been proposed for high school during the creation of this document are the following:

1. *Demographics:* statistics; prediction; variation; connection to society; rates of change; and unusual types of graphs and charts to create and interpret

2. *Geodesics:* shortest distances on different surfaces, especially cones, cubes, and spheres; spider-and-fly; measurement; variation; geometry; and multiple representation
3. *Scheduling and distribution:* allocation of scarce resources in material and time; fair distribution; sensitivity to perturbations; classification; and mathematical bases for decision making

The previous units explore subject matter whose scope is relatively small; that is, they will probably not be visited more than once. They also highlight areas of mathematics not traditionally treated in high school mathematics.

Instructional Materials Criteria for Kindergarten Through Grade Eight

A p p e n d i x **A**

Instructional materials, which provide the base of a mathematics program, determine which mathematical ideas students encounter. In addition, they affect how students (1) interact with mathematical ideas; (2) develop attitudes toward mathematics; and (3) make sense of mathematics. Instructional materials also greatly influence the way a teacher teaches; that is, what the teacher does in the classroom that supports the students' learning.

The instructional materials criteria describe an ideal mathematics program that is aligned with the goals and program characteristics identified in this *Framework*. These criteria are designed to be used in the review and evaluation of basic instructional materials in mathematics.[1] The criteria are organized in the following categories:

1. *Mathematical content*—what mathematical ideas and subject matter provide the basis for the instructional program
2. *Program organization and structure*—how the program is organized in a year's work of cohesive units, lessons, and tasks
3. *The work students do*—what the students work on and how they do it
4. *Student diversity*—how the program deals with diverse cultural and economic backgrounds, achievement levels, English-language proficiencies, and interests of students
5. *Assessment*—how the program integrates assessment with instruction and helps teachers assess student performance
6. *Support for the teacher*—how the materials support what the teacher does in the classroom

Each category offers a different perspective on how the total program is intended to be experienced by the students. In using these categories, reviewers should keep in mind the following general points:

1. These categories are not distinct; they overlap. For example, students' experiences in a program cannot be accurately judged simply by looking at the category titled "The Work Students Do." Their experiences will also be affected by the quality of the teaching, the kinds of units and tasks the students work on, and the mathematical content of the program.
2. In each category all of the components of an instructional program (such as student materials, teacher materials, and technology) are to be examined as to how well they work together to provide a quality program for students in a classroom. These criteria do not presuppose the presence or absence of any particular component. It is possible to design (a) a complete program that does *not* have a single student textbook; and (b) a high-quality program that has

---

[1]The criteria are designed to be used with complete programs (at least a full grade level) rather than with supplemental materials.

a student textbook at its center. Similarly, videotapes, computer software, and other technology might or might not be included in a program.

3. Within each category is a series of subpoints. These subpoints are not necessarily of equal weight and should not be judged individually. Instead, they should be used as an aid to identifying the qualities that contribute to a category.

4. Most students and teachers have not experienced the kind of mathematics programs called for in this *Framework*. Therefore, reviewers using these criteria must carefully evaluate how effective the instructional materials will be in the classroom. Instructional materials need to be descriptive enough to help conscientious teachers implement a new kind of program yet not so tightly structured that teachers have little flexibility in implementation.[2]

Mathematical Content

Important mathematical ideas are at the heart of the program. Students learn and use concepts or skills within larger mathematical inquiries and visit the central and most useful mathematical ideas again and again in different contexts. As students mature, they investigate those ideas more and more deeply. The program reflects the *Framework*'s specifications for mathematical content:

- The strands, as defined in the 1985 *Mathematics Framework* and updated in this *Framework* and in the NCTM *Standards*, are incorporated throughout the program. Other strands not delineated in these documents may also appear if the mathematics in the eight strands of *functions, algebra, geometry, statistics and probability, discrete mathematics, measurement, number, and logic and language* is developed. Units of instruction typically interweave ideas from more

---

[2]It should not be assumed that instructional materials are solely or primarily responsible for providing a quality mathematics program. This *Framework* recognizes the necessity of providing professional development programs for teachers and redesigning national, state, and local assessment practices. Nonetheless, alignment of instructional materials with this *Framework* is critical.

than one strand, and a year's work incorporates a balanced treatment of all of the strands.

- The unifying ideas are fully explored over the course of a given year. In addition to the unifying ideas presented in this *Framework*, which must be developed, a small number of other unifying ideas may be added. In particular, within the span of kindergarten through grade five, emphasis is given to the unifying ideas of (a) *quantifying;* (b) *finding, making, and describing patterns; and* (c) *representing quantities and shapes.* And within the span of grades six through eight, emphasis is given to the unifying ideas of (a) *proportional relationships;* (b) *multiple representations;* and (c) *patterns and generalization.*
- Subject matter for units is drawn from coherent clusters of interrelated ideas. Although no single unit topic from Chapter 4 is required, the units used in the program are consistent with the description of units on pages 89–96.

In addition, number sense is developed as students encounter and use quantities and their representations in realistic contexts. Students are enabled to:

- Devise their own procedures for problems that involve calculations through their continuing work to produce numerical results for a purpose.
- Decide consistently on the most efficient means for calculating a numerical result in a given situation with the mind only or with the aid of paper and pencil or a calculator.
- Use realistic numbers in problems. Problems in upper elementary grades and beyond have not been contrived to confine solutions to integers.

Program Organization and Structure

The instructional materials present students with coherent, connected, and accessible mathematical experiences organized in a year's or several years' work of units, lessons, and tasks. Units are cohesive chunks of work lasting one to six weeks. A unit consists of lessons of one or more related tasks that students work on. A year's work may include some activities that do not belong to any unit, such as a few excellent but otherwise disconnected lessons, and especially time taken out to reflect on mathematical connections between units. The instructional materials describe how teachers can present the tasks to the students, how students are to work on the tasks, and how students are expected to reflect on and share their work with the rest of the class.

The program is organized and structured as follows:

- The basic building block of the program is a unit, a collection of activities with instructional and mathematical coherence. Units include not only the development of mathematical ideas but also their use in many settings and their integration with previous learning.

- Each year's work includes approximately ten units, with a balance among strands and substantial attention to each unifying idea. In the primary grades units will be shorter and more numerous.

- Assignments vary in length. Some, such as investigations, may be quite large and, especially in later grades, form the backbone of the unit. Other tasks may be more limited in scope and provide the tools or techniques for successful completion of the investigation.

- Units include time for reflecting and summarizing the material within the unit.

- Some units may be distributed over rather than concentrated within a period of time, with activities from these units interspersed within or between other units.

- Although most activities, including exercises and problems, are within the boundaries of a coherent unit and related to it, some activities may be outside a unit structure.

- Some tasks are quantitative, interdisciplinary, "real-life" problems; others are more purely mathematical investigations, including games and puzzles.

- Lessons and tasks within a unit frequently involve concepts from more than one strand and explore their interconnections and relationships. Students encounter mathematical ideas in different ways as well as in different contexts.

- Some tasks may be incorporated in a menu format whereby students decide in what order the tasks will be done; or the tasks may be sequenced.

- The instructional materials outline a default sequence for the units and indicate how teachers can use the program flexibly by rearranging, substituting, or modifying units and tasks to meet the needs and interests of their students.
- From year to year the units and the tasks within them increase in depth and complexity. Students encounter the same unifying mathematical ideas in different or more complex contexts. Gradually, tasks increase in duration and include more abstraction and formalism. Students grapple with increasingly complex questions and investigations and assume more and more responsibility for developing complete and comprehensive reports or products.

The Work Students Do

Students, who are expected to think and reason in all their mathematical work, are presented with a variety of challenging mathematical tasks and investigations they find meaningful. The instructional materials engage students in experiences that require the students to think mathematically and convey to the students that at times they will be uncertain. The students are frequently asked to conjecture and pursue possibilities without knowing whether they will obtain an answer. The work students do in the program incorporates the following:

- Assignments help students develop all four dimensions of mathematical power. That is, students are asked to think and communicate, draw on mathematical ideas, and use mathematical tools and techniques effectively.
- Students encounter a varied program, including all of the strands and unifying ideas, and a balance among exercises, problems, and investigations. Many assignments are open-ended and allow multiple approaches.
- Students experience both challenge and success. Some tasks require time and deliberation and are continued over several days. Materials help teachers give students a clear standard of performance and suggest ways to help students meet the standard.

- Students are asked to formulate mathematical questions and assess what is known and what must be determined.
- Students generally choose the approaches to take, the depth to pursue, the reporting format to use, and the manner in which to extend a task.
- Students are frequently asked to interact with one another and often work in small heterogeneous groups. They are expected to share approaches, conjectures, difficulties, results, and evidence within their group and with other groups.
- Students are asked directly to reflect on the work they are doing and the mathematical understandings they are developing.
- Students are asked to propose and test generalizations as they become apparent and make connections among the mathematical ideas within a lesson or among lessons.
- Students use manipulative materials and technology to explore ideas and solve problems. A variety of tools is continually available for students to use.
- Every student has access to a calculator at all times for use in class and outside and may choose to use it for any occasion except when attempting to develop mental dexterity.
- Students are consistently asked to communicate their findings orally or in writing. At times students are asked to present their results and conclusions to various audiences, such as an expert, a layperson, a parent, a teacher, a peer, a younger sibling, or a neighbor, to inform, explain, teach, or persuade.
- Students are asked to explore a situation, gather data, or interact with members of their families for homework assignments.

Student Diversity

The materials recognize that students have a wide range of cultural and economic backgrounds as well as achievement levels and interests. All students participate fully in each unit.

Assistance provided to students having difficulty is in addition to and not instead of the regular program, with the goal of supporting successful participation in the regular program. The materials reflect the assumption that a number of students in a class, not all necessarily with the same linguistic background, have limited proficiency in English. However, the program is not limited to computation or otherwise diluted for such students.

The instructional materials provide for student diversity in the following ways:

- The tasks and problems students work on are accessible to all students. They are rich and open and can be investigated at many different levels.
- Many tasks or lessons use students' personal, family, or cultural experiences to create the specific context for the lesson.
- The materials provide a variety of suggestions for teachers of students whose primary language is not English. Included are suggestions for students' use of their primary language to the extent possible and techniques for sheltering the presentation for students whose proficiency in English is sufficient for them to be successful in sheltered classes.
- Students frequently work in collaborative learning groups. In particular, students are encouraged to value the points of view and experiences of others.
- The materials advise teachers on how to deal with aspects of lessons that might generate conflicting feelings or responses from students. The materials include suggestions that promote sensitivity and respect for each student.
- There is room in each unit for some students to go more deeply into some aspect of the unit's investigations according to their interests or their rapid grasp of the ideas.
- The materials advise teachers on the use of appropriate strategies, peer support, and, if available, instructional aides, to support the learning of all students in the class.

Assessment

Assessment is integrated in the instructional program. The instructional materials help teachers use assessment in a variety of ways to get information about what individual students or groups of students understand and are able to do in solving mathematical problems. Assessment is built into the program in such a way that:

- Specific assessment tasks are included in the units. Most have the same character as learning tasks. That is, students formulate problems, consider and apply a variety of approaches, and determine and explain their findings.
- Students use tools, such as calculators and manipulatives, and other resources, such as notes or reference materials, while they are carrying out assessment tasks.
- Students have ample time to work on assessment tasks. They frequently have the opportunity to revise and resubmit important assignments to bring their performance up to high-quality standards.

The materials include suggestions to the teacher concerning how to:

- Use learning tasks for assessment purposes.
- Observe, listen to, and question students while they work and how to keep track of insights about the students they may have.
- Organize and use student portfolios.
- Keep parents informed about the progress of their children and the variety of assessment methods being used—what they are and why they are important.[3]
- Involve students in self-assessment.

Support for the Teacher

The instructional materials describe the mathematics program so that teachers can implement it. The materials provide specific suggestions and illustrative examples of how the teacher can facilitate the student behaviors identified under "The Work Students Do." In providing support for the teacher, instructional materials include the following:

---

[3]Information on student progress may include ways of communicating about students' work, using samples of students' work (from portfolios) in parent conferences, having students talk frequently with their parents about their work, and so on.

- Description of the important mathematical ideas in the units
- Discussion of how students are to encounter the mathematical ideas and how the experiences are related to what is known about children's learning or developmental level
- Descriptions or pictures of what units and lessons will look like when implemented in the classroom
- Information on what is important to do and say in a lesson or unit
- Suggestions on questions to ask and ways to respond that keep students' thinking open and help students reflect on what they have done
- Suggestions on how teachers might think about and reflect on what happens in a lesson or unit
- Suggestions on when to present information to students and how to do so
- Suggestions on working with a diverse classroom of students (See "Student Diversity.")
- Suggestions on helping students work together productively
- Suggestions on managing manipulative materials, calculators, computers, and other tools so that they are accessible when the students want to use them
- Suggestions on helping students write more effectively about their mathematical thinking
- Suggestions on involving parents and keeping them informed about the program, including sample letters that communicate the values of the program and the rationale for what students are doing in the classroom
- Suggestions on assessing student performance (See "Assessment.")

Coherence
Across the
Elementary
Grades

A p p e n d i x **B**

What follows are sample experiences that students might have at different grade levels. The reader is cautioned that these items represent a sampler, not a syllabus, and are neither required nor comprehensive. They serve only to clarify the material in the section titled "Suggested Units for the Elementary Grades," pages 109–16.

	Kindergarten and grade one	Grades two and three	Grades four and five

Attributes and classification ▶

Children in kindergarten and grade one <u>identify</u> <u>attributes</u> of things they can see as they <u>physically</u> <u>sort the objects in a variety</u> of ways. For example, when looking at a collection of buttons, they may state that "<u>some have two</u> <u>holes</u>," "<u>some are white</u>," and "<u>some are metal</u>." During early sorting experiences students sort a group into two piles—those that have the attribute and those that do not. Children at this age level may have difficulty understanding how an object can simultaneously fit into two overlapping categories, such as buttons that have two holes and are made of metal. Students make collections of things that go together; <u>play guessing</u> <u>games to sort collections</u> ("This goes into the pile of objects I'm thinking of. What else does?"), or identify a specific object in a collection ("No, the object I'm thinking of is not pink.").

Students begin to classify things by two (and, later, three) overlapping attributes. For example, students may decide to overlap yarn circles (Venn diagram with one intersection) when they are sorting students into those wearing watches and those with brown eyes. They discuss fuzzy attributes, which have no clear-cut boundaries, such as tall/short or brown hair. They discuss the need for clarity and precision when specifying attributes. They solve student-constructed sorting puzzles by guessing the attribute or the rule. They use clues for solving riddles about tiles, money, shapes, or other objects. The students develop ways of recording and communicating their classification system.

Often, as part of their work on other units of study, students classify many things. They classify numbers (e.g., multiples, primes, composites, factors, odd or even) and geometric shapes (e.g., squares are also rectangles; angles are larger or smaller than a right angle). They use various forms of representation, such as Venn diagrams, to communicate their work to others. They write riddles with clues for other students to solve (possibly using geometric concepts, number properties, or number of coins). They examine each other's riddles for clarity, redundancy, or ambiguity.

186

Kindergarten and grade one	Grades two and three	Grades four and five	

Students have many experiences to develop the notion of a number. They count objects and arrange objects to represent an amount concretely. They show that there are various ways to arrange a number of objects and that a number may have several names. For example, to illustrate *fiveness*, students arrange five objects in groups of four and one, two and two and one, and so on. They compare different groups of objects, discussing whether they are the same amount or whether one group has more or fewer objects than another group. They count by ones, upward and backward, and skip count (by twos). They cut fruit (or other objects) into halves or fourths and share so that each person has the same amount.

Students count, compare, order, and estimate larger quantities. As they work with larger quantities, they group objects and investigate place value. They use fractions and decimals encountered in everyday situations. The students find halves (and other common fractions) of wholes and of groups. They recognize that a half dollar is 50 cents, can be written as $.50, and may show on their calculator as 0.5 (e.g., when they have divided $1 by 2).

Students extend their understanding of place value ideas to make sense of large numbers and decimals. They interpret numbers encountered in the media and in their everyday experiences. For example, they devise ways of explaining how much a million of something is. They begin to use, in natural ways, common fractions (such as halves, fourths, and tenths) interchangeably with their decimal and percent equivalents. For example, when interpreting an advertisement for a 25 percent off sale, a student might say that the prices are 1/4 off; and for a 35 percent off sale, the prices are a little more than 1/3 off. The students explore different ways of categorizing and representing numbers and so on. They informally investigate primes and composites, square numbers, odd and even numbers, and so on. They look for patterns, some numerical and/or geometrical, to describe a category and predict which numbers belong in it.

◄ **Understanding number and numeration**

187

Kindergarten and grade one	Grades two and three	Grades four and five

Understanding arithmetic operations ▶

Students begin informally to use ideas of operations to answer questions in everyday situations as an extension of counting: "How many books are there if Juan has five and Jody has seven?" "If there are 25 students enrolled in our class and three are absent today, how many are here?" "Will there be enough candy for three apiece if we share this bag?" "If everybody brings two cans of food, how many will our class have for the Thanksgiving Food Drive?" "If I have 35 cents, how much more money do I need to have $1?" Students solve problems like these experientially: count, use materials, act out the situations, discuss their thoughts with each other and the teacher— doing things that make sense to them in light of the particular situation. They use number sentences as a way to record their experiences. From their many experiences using numbers, they recall some of the addition/ subtraction facts.

Students continue to solve problems they encounter in their everyday experiences, making sense of each situation by acting out the situations, drawing pictures or diagrams, and using materials. They explore and share how different number sentences can model the same situation. For example, in addressing the question "How much money does Sara need to save to have $10 if she already has $6.45?" students respond that the problem could be modeled by $10 − $6.45 = ? or $6.45 + ? = $10 and could be solved by subtracting or counting on. Students investigate different ways in which they can use an operation. For example, they use multiplication in a variety of settings: finding the cost of six pencils at 12 cents each; studying areas by placing tiles on the bottom of boxes; building identical block towers from a set number of blocks (e.g., 20). They learn the addition and subtraction facts and some of the multiplication and division facts through their many encounters with numbers.

Students solve problems involving more complex situations. For example, they might determine the number of pizzas they need to sell at $4.25 each, with a profit of $.75 per pizza, to earn enough money to send the class of 28 students on a field trip that will cost $5.95 per student. They share the procedures they use, discussing why different approaches produce the same numerical result. They calculate without assistance or with the aid of paper and pencil or a calculator, determining which method seems most appropriate for a given situation and judging whether an approximation would suffice rather than calculating an exact answer. They extend their understanding to explain operations with rational numbers and powers of ten. For example, they explain or illustrate with a situation or picture what $\frac{1}{2} \times \frac{1}{3}$ means and why affixing zeros is a shortcut to multiplying by powers of ten.

Kindergarten and grade one	Grades two and three	Grades four and five	

Students' first experiences with data develop from their curiosity about classmates and everyday events. "Which snack do you choose, an apple or an orange?" "Who is the tallest?" "How many pets do you have?" "Do you have a dog? Cat? Bird?" To find out, students count and measure. For example, they line up their snack choices to see whether more children chose an apple or an orange. They order themselves by height. They compare the number of pets each student has and may tally up the total number for the class. They classify pets by kind. They record their findings with pictures and simple graphs.

Students also deal with data as part of classroom routines. For example, as a way of taking roll, they each day turn over their pictures on an attendance graph. A quick glance tells them who is absent (or forgot the routine). Periodically, they may also use these pictures to record data on other graphs, such as one titled "How Did You Get to School Today? Walk or Ride?"

As a class students routinely graph many things and summarize the results. Students do simple surveys of their class-mates, families, or neighbors. Then they discuss and decide on categories for organizing this information. For example, a class may decide to find out what games people like to play. After questioning a variety of people, the class may decide to place the games in categories such as ball games, card games, outdoor games, or thinking games. To do so, they discuss issues such as what kind of categories make sense and how some categories overlap. They explore different ways of displaying the data, such as using Venn diagrams, graphs, and tables.

Working by themselves, with a partner, or in small groups, students research a question of interest, collect data, organize them, display them in graphs, and interpret the results. They examine data from the media and speculate on how the findings were determined. They discuss different ways of collecting information. They conduct sampling experiments (e.g., with cubes and beans in a bag) and compare results from a number of samples with the entire population of cubes and beans. They survey a random sample of students at their school on a question of interest and then compare these results with a written survey of the entire school.

◀ **Dealing with data**

Kindergarten and grade one	Grades two and three	Grades four and five

The process of measurement ▶

At this grade span the process of measurement is very exploratory. Students directly compare two objects to determine, for example, which of two children is taller. They count the number of steps across the room or the number of scoops it takes to fill a jar. They begin to measure length with nonstandard units by using a number of the units end to end. For example, they use unifix cubes snapped together to measure the length of the tray on an easel or a number of unsharpened pencils end to end to measure the length of the chalkboard tray. (Students may want to discuss why the measure is different if used pencils are employed to measure the chalkboard tray.)

Students use a variety of nonstandard measuring units to measure the length of different things. They discuss how the size of the object affects their choice of measurement unit and how some choices are more appropriate because of the size of the object. For example, they might discuss whether it makes more sense to measure the length of a table by using toothpicks or straws or counting paces. They compare using one straw to using many. They discuss the importance of accurately keeping track of endpoints when iterating with the straw and keeping an accurate count. As students compare the results they get using a unit that varies in size, such as a pace or hand span, they discuss the need for standard-sized units. They informally begin to use common standard units and tools.

Students also continue to explore measures other than length (e.g., area, volume, mass/weight, temperature), using nonstandard and standard units. For example, they use pan balances to compare the weights of different objects. They practice "reading" measuring instruments, such as rulers, scales, and thermometers, and discuss what the measures mean.

Students use standard measurement units and tools to measure many items. They choose the units and tools and compare results. They develop referents for common standard units and use these referents to judge whether measurements are reasonable. For example, a student may think of a meter as about the width of a door and, therefore, cannot make sense of a ten-meter running race. They discuss the accuracy needed for different measurement tasks.

190

Kindergarten and grade one	Grades two and three	Grades four and five	
Activities are identical with those described under "Exploring Process of Measurement." The same activities lay the foundation for both areas of study.	Using a variety of non-standard measurement units, students measure the linear dimensions of different objects. They explore the concept of area by seeing how many books will cover the top of the table or whether all the students' drawings will fit on the bulletin board. They explore the concept of perimeter by determining how long the border needs to be to frame the bulletin board. They explore the concept of volume by filling boxes with blocks or bottles with rice. Frequently, they compare the sizes and order the objects they have measured.	Students determine and compare areas and perimeters of rectangular and nonrectangular shapes. (A few students may invent workable formulas for area and perimeter.) They draw different shapes that have the same area and/or perimeter. They explore the effect of changing in the length of one or both dimensions of a rectangular shape on its area and/or perimeter. They explore the relationship between area and perimeter. They also continue some informal work with volume.	◀ **Measuring geometric figures**

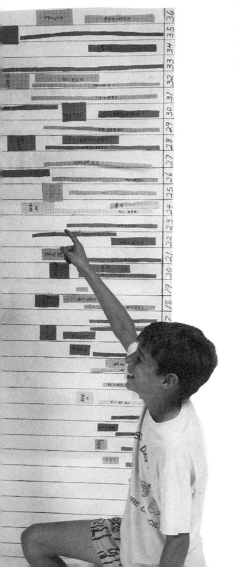

Kindergarten and grade one	Grades two and three	Grades four and five

Locating and mapping ▶

Children naturally develop locational vocabulary as they play and talk about their lives. "The lamp is *on* the table." "The ball is *under* the chair." "Joy lives *across* the street." "Grandma lives *far away.*" They model the world when they play with a doll house or build a race course for their model cars. They tour the school, locating the office, library, and so on, and share various routes to the same location. They take a neighborhood walk and visit students' houses, parents' businesses, and so on. They informally compare distances by deciding who lives the farthest from school or which classroom is the closest to the playground. Or they share the number of steps from their classroom door to the teacher's desk.

Students give "Logo-like" directions to move a "robot" (person) around the room. They measure distances to improve direction-giving. They begin to estimate distances (using familiar units) to various locations in the room and in the school. They write directions about how to get from one place in the school to another or how to get from school to their home and model the directions on maps. They play Copy My Desktop by placing a number of items on the top of a desk and drawing a bird's-eye perspective map of the desktop. Students exchange maps and recreate other students' desktops. They model their classroom, using blocks or other items and draw a floor plan of their bedroom or classroom, discussing whether they fit into the model or drawing.

Students make enlarged or reduced-sized drawings or models. They begin to interpret scale drawings and models. The students use maps of the city, state, or U.S. to locate places and determine distances. They make maps and write directions to "buried treasures" and follow each others' maps and directions. They discuss advantages and disadvantages for various methods of telling or showing others how to get from place to place.

			◄ Visualizing and representing shapes
Students begin to visualize, describe, and represent shapes when they informally play with blocks (both 3-D "kindergarten blocks" and 2-D design blocks): "I made a steep hill for my toy car." "I'm going to make a design just like yours. Let me see; I need to put a green block here and then a red one." "I made a house for my picture by using an orange square with a green triangle on top." Later, students use toothpicks, tiles, cubes, blocks, and other materials to construct two- and three-dimensional shapes and figures. They often make tracings or drawings to record what they have done. They look for shapes in the real world. The emphasis is on the exploratory; vocabulary develops naturally as students use words that the teacher has used to describe shapes.	Students become aware of the properties of shapes through many experiences sorting and classifying shapes. How many sides? How many faces? Which shapes have curves? Which shapes look the same upside down? Which shapes have pointed corners? How are the shapes alike? Different? They build shapes by putting other shapes together and cut shapes in half to see what shapes are formed. They play What's Behind My Wall? by building a pattern or design with a few blocks and then describing it so that another student is able to reproduce it.	Students identify and describe properties of shapes and relationships among them. For example, they classify quadrilaterals, noticing that squares are also rectangles and parallelograms. They explore the notions of symmetry, congruence, similarity transformations (slides, rotations, and reflections). For example, they complete a pattern as it would look if reflected in a mirror placed on a specified line. The students investigate the relationships between three-dimensional objects and two-dimensional representations by making networks ("jackets") for blocks and building block models after examining several views (front, side, top, and so on) of the structure.	

193

Kindergarten and grade one	Grades two and three	Grades four and five

Exchange ▶

Kindergarten students have a variety of experiences with exchange. They buy things with money. They sometimes trade items with each other (such as sandwiches). They play store, counting out money for purchases. They trade five pennies for a nickel and ten pennies or two nickels for a dime. They do geometric puzzles and designs with pattern blocks or tangrams, seeing that different shapes can be exchanged to fill the same outline. They discuss whether exchanges are fair or equivalent (e.g., "Do we each have the same amount if you have one-half of an apple and I have two-fourths? What if you have one-half of an apple and I have two-fourths of an orange?").

Students solve a variety of problems involving the exchange of money. They "buy" different items: They select the items, calculate the total cost, "pay" for them, and practice making change. They find all the different ways to show a particular amount of money and record their results in a table, picture, or chart. They choose something they want to buy and write a plan for how they will save for it. They compare the price of an item at different stores and during sales.

Students use money and manipulatives to explore place value concepts. They group objects in multiples of ten or exchange ten objects for one worth ten times the value.

Students investigate such questions as what does a typical lunch (brought from home) cost? Or how much does a school outfit cost? Working in small groups, students determine what makes a lunch or an outfit typical. Then they shop around, researching the prices of component parts of the lunch or outfit at different stores and calculating total costs, including tax. Each group presents a report of what it did and its findings, including the factors that influenced the choices it made. For example, how important was style, quality, or price? Students make sense of percents in sales tax and in percent-off sales.

Students determine or substitute equivalent measurements in making recipes and building models. They determine the amounts of ingredients needed for making a multiple recipe (or a portion of a recipe). They explore the concept of scale (e.g., see on a map how an inch represents several miles or make drawings of common objects that are twice [or half] as large as the objects).

Kindergarten and grade one	Grades two and three	Grades four and five	

Students learn rules and play games. They match pairs of cards and spin a spinner to see who gets the highest number. They play lotto games.

They practice taking turns and waiting for a turn. They discuss the fairness of situations that occur in the classroom and on the playground.

Students play thinking games such as Guess My Number and Guess My Word. They discuss strategies for playing a game.

Students begin probability activities. They play Cubes in a Bag. They create simple board games and games involving cards and spinners.

Students begin to categorize games and discuss how the games are different and how they are similar.

Students have many experiences with probability in the contexts of diverse games. They spin spinners, throw dice, and record the results. They decide whether or not games are fair by investigating, most often empirically, the probability of winning.

Students identify how luck and skill are involved in various games. They invent more sophisticated strategies, design more complex games, and write rules for them. They change their rules and the rules of other games, discussing what effect the changes have on the play of the game.

◀ **Games and rules**

195

Traditional, Alternative, and Desired Classroom Practices

A p p e n d i x **C**

Appendix C contains lists of traditional, alternative, and desired practices related to classroom instruction in mathematics. *Note*: The middle column, "Some Alternative Practices," is not an intermediate or necessary step; it illustrates what some teachers have been doing since the 1985 *Framework* was published.

	Traditional practices	Some alternative practices	Desired practices
Lessons ▶	A two-page lesson per day is organized according to a specific objective. Lesson includes: • Application example • Instruction in how to do a specific procedure, perhaps with some simplified explanations • A few exercises to check for understanding • Practice exercises • Optional enrichment problem	Lesson lasts for one day or two days. Work sheets provide specific directions for the student on what to do and how to record results.	Most lessons last for several days and generally involve more than one big mathematical idea. Students work with important ideas over an extended, continuous period, sometimes as long as six weeks.
Assignments ▶	Daily assignments are given from the textbook and are expected to be done individually. Drill and practice assignments from the textbooks are given frequently.	Students use manipulatives and work in small groups, but fairly prescriptive work sheets often guide students' work. There may even be a project, but the mathematics behind it may not be evident from the product.	Assignments develop students' mathematical power; they are challenging and multidimensional. They may extend over several days and require considerable time outside the class. Often, they are cast in the form of a broad, complex problem to investigate as a group. Findings or results may be presented in an individual or group report, an oral presentation, visual displays, or all three.

Traditional practices	Some alternative practices	Desired practices	
Procedures are given major focus. Emphasis is placed on learning the steps to perform an algorithm and on providing enough practice so that the procedure becomes automatic for students.	Procedures are given major focus. Manipulatives are used to teach and explain the algorithm, although students are still expected to become proficient performing the algorithm with paper and pencil so that they can do well on tests that require quick and accurate computation.	Students develop a range of computational procedures, with greater emphasis on number sense than on algorithms. They invent and use a number of different computational procedures as aids to solving problems. Students are expected to decide the most efficient means of calculating an answer for a given situation; that is, to work on the problem with the mind only or with the aid of paper and pencil or a calculator.	◀ **Computational procedures**
Manipulatives are used in primary grades at the teacher's discretion and are not part of the students' textbooks.	Manipulatives are used frequently but in prescribed ways to teach specific concepts or procedures.	Students have manipulatives available at all times. Sometimes students use manipulatives for specific purposes. At other times the choice of manipulatives is up to the students.	◀ **Manipulatives**
Calculators are used only for a few calculator lessons and are not permitted for use on daily assignments and tests.	Calculators are used for calculator lessons and problem-solving lessons. In some programs students use calculators to check paper-and-pencil computations. In others students use calculators after paper-and-pencil algorithms are mastered. Students are not permitted to use calculators on tests.	Students have calculators continually available for use: in class, on homework assignments, and on tests. Students conduct experiments based on using calculators to explore the behavior of mathematical structures.	◀ **Calculators**

Traditional practices	Some alternative practices	Desired practices

Computers and software ▶

Computers are used for drill and practice, if at all, and then often in a computer lab staffed by a specialist or aide. The computer helps the teacher only as a grade book.

Computers are used more frequently throughout the school and in the classroom. Students use a few imaginative pieces of software that usually focus on specific types of problems. Computer work is considered enrichment and is not connected to students' core work. Students work in groups because there are not enough computers.

Computers are available at all times. Students may use computers by themselves or in groups. Teachers and students use more sophisticated special-purpose programs and are fluent enough with tool software, such as word processors, spreadsheets, and data bases, to use the computer to meet their particular needs. Computers are viewed as extensions of human capability and not ends in themselves.

Student interaction ▶

Most of the time students are expected to work alone. They sit at individual desks, facing the front of the room. But primary grade students may sit at tables.

Most of the time students are expected to work alone. For problem-solving activities, students move their desks into groups of four.

Students sit in small heterogeneous groups. They are encouraged to interact with each other. For many assignments students may choose whether to work alone or with others.

Traditional practices	Some alternative practices	Desired practices	
Students are placed in ability-level groups (elementary) or tracks (secondary), mainly on the basis of how quickly they can perform paper-and-pencil computations or algebraic manipulations. In elementary school, students often work in the same class and study the same material but at different rates. In secondary school, students are grouped in separate classes, with different titles, and study vastly different materials.	In elementary school the teacher teaches the heterogeneous class as a whole group but pulls out and gives additional assistance to students who are having difficulties. Some secondary schools institute Math A classes. Secondary teachers discuss tracking problems but see no workable alternative to the traditional college preparatory courses.	All students study a common core curriculum. In kindergarten through grade eight, classes are grouped heterogeneously. Students with special interests or talents go more deeply into some investigations, often working with an interest group. Students having difficulty sometimes get additional assistance, often before or after school. Some students begin the high school core sequence early, most in ninth grade, and the remainder after completing Math A and, perhaps, Math B.	◄ **Grouping**
Approximately the first third of the year is spent in reviewing computation and symbolic manipulation topics that have been taught previously. New material is presented in chapters but is reviewed only on semester tests.	Review is provided during ten-minute warm-up sessions at the beginning of the period or in "sponge" activities done sporadically. Occasionally, students work on problems that incorporate work they have studied from several strands.	Review is built into the rich and complex problems students work on.	◄ **Review**

Traditional practices	Some alternative practices	Desired practices

Tests and assessment ▶

Testing is considered important. The program may use the test-teach-test cycle. Tests closely match assignments, with a narrow focus on skills. Questions have one correct answer. Tests can be scored objectively. Student mastery is expected. The goal of testing is to assess and classify the student.

Formal tests are important. Although teachers question their usefulness and impact, they expect their students to do well on norm-referenced tests. Some teachers include open-ended questions and require students to explain their thinking on their tests.

The goal of assessment is to evaluate student work rather than the student. There are many forms of external and internal assessment, including portfolios of student work, observation, interviews, and group work. Students are expected to explain their thinking. Holistic scoring rubrics focus on all dimensions of mathematical power. Often, students are expected to revise their work to meet the standards. Students participate in assessment.

Calculus
in the
Secondary
School

A p p e n d i x **D**

This 1986 letter from the Mathematics Association of America (MAA) and the National Council of Teachers of Mathematics (NCTM), addressed to secondary school mathematics teachers, concerns instruction in calculus in the secondary school.

Dear Colleagues:

A single variable calculus course is now well established in the twelfth grade in many secondary schools, and the number of students enrolling is increasing substantially each year. In this letter we would like to discuss two problems that have emerged.

The first problem concerns the relationship between the calculus offered in high school and the succeeding calculus courses in college. *The Mathematical Association of America (MAA) and the National Council of Teachers of Mathematics (NCTM) recommend that the calculus course offered in the twelfth grade should be treated as a college-level course.* The expectation should be that a substantial majority of the students taking the course will master the material and will not then repeat the subject upon entrance to college. Too many students now view their twelfth grade calculus course as an introduction to calculus with the expectation of repeating the material in college. This causes an undesirable attitude on the part of the student both in secondary school and in college. In secondary school all too often a student may feel, "I don't have to master this material now because I can repeat it later"; and in college, "I don't have to study this subject too seriously because I have already seen most of the ideas." Such students typically have considerable difficulty later on as they proceed further into the subject matter.

MAA and NCTM recommend that all students taking calculus in secondary school who are performing satisfactorily in the course should expect to place out of the comparable college calculus course. Therefore, to verify appropriate placement upon entrance to college, students should either take one of the advanced placement (AP) calculus examinations of the College Board or take a locally administered college placement examination in calculus. Satisfactory performance on an AP examination carries with it college credit at most universities.

A second problem concerns preparation for the calculus course. *MAA and NCTM recommend that students who enroll in a calculus course in secondary school should have demonstrated mastery of algebra, geometry, trigonometry, and coordinate geometry.* This means that students should have at least four full years of mathematical preparation beginning with the first course in algebra. The advanced topics in algebra, trigonometry, analytic geometry, complex numbers, and elementary functions studied in depth during the fourth year of preparation are critically important for students' later courses in mathematics.

It is important to note that at present many well-prepared students take calculus in the twelfth grade, place out of the comparable course in college, and do well in succeeding college courses. Currently, the two most common methods for preparing students for a college-level calculus course in the twelfth grade and to begin the first algebra course in the eighth grade or to require students to take second-year algebra and geometry concurrently. Students who begin with algebra in the ninth grade who take only one mathematics course each year in secondary school should not expect to take calculus in the twelfth grade. Instead, they should use the twelfth grade to prepare themselves fully for calculus as freshmen in college.

We offer these recommendations in an attempt to strengthen the calculus program in secondary schools. They are not meant to discourage the teaching of college-level calculus in the twelfth grade to strongly prepared students.

Changes in Content and Emphasis— NCTM *Standards*

A p p e n d i x **E**

The following summaries are taken from the NCTM *Standards* (K–4, pages 20–21; 5–8, pages 70–73; and 9–12, pages 126–29). The reader is advised to refer to the *Standards* for a complete discussion.

Summary of Changes in Content and Emphasis for Mathematics in Kindergarten Through Grade Four

Increased attention	Decreased attention
Number ▶ • Number sense • Place-value concepts • Meaning of fractions and decimals • Estimation of quantities	• Early attention to reading, writing, and ordering numbers symbolically
Operations and Computation ▶ • Meaning of operations • Operation sense • Mental computation • Estimation and the reasonableness of answers • Selection of an appropriate computational method • Use of calculators for complex computation • Thinking strategies for basic facts	• Complex paper-and-pencil computations • Isolated treatment of paper-and-pencil computations • Addition and subtraction without renaming • Isolated treatment of division facts • Long division • Long division without remainders • Paper-and-pencil fraction computation • Use of rounding to estimate

Increased attention	Decreased attention	
• Properties of geometric figures • Geometric relationships • Spatial sense • Process of measuring • Concepts related to units of measurement • Actual measuring • Estimation of measurements • Use of measurement and geometry ideas throughout the curriculum	• Primary focus on naming geometric figures • Memorization of equivalencies between units of measurement	◀ **Geometry and Measurement**
• Collection and organization of data • Exploration of chance		◀ **Probability and Statistics**
• Pattern recognition and description • Use of variables to express relationships		◀ **Patterns and Relationships**
• Word problems with a variety of structures • Use of everyday problems • Applications • Study of patterns and relationships • Problem-solving strategies	• Use of clue words to determine which operation to use	◀ **Problem Solving**

	Increased attention	**Decreased attention**
Instructional Practices ▶	• Use of manipulative materials	• Rote practice
	• Cooperative work	• Rote memorization of rules
	• Discussion of mathematics	• One answer and one method
	• Questioning	• Use of work sheets
	• Justification of thinking	• Written practice
	• Writing about mathematics	• Teaching by telling
	• Problem-solving approach to instruction	
	• Content integration	
	• Use of calculators and computers	

Summary of Changes in Content and Emphasis for Mathematics in Grades Five Through Eight

Problem Solving ▶	• Pursuing open-ended problems and extended problem-solving projects	• Practicing routine, one-step problems
	• Investigating and formulating questions from problem situations	• Practicing problems categorized by types (e.g., coin problems, age problems)
	• Representing situations verbally, numerically, graphically, geometrically, or symbolically	
Communication ▶	• Discussing, writing, reading, and listening to mathematical ideas	• Doing fill-in-the-blank work sheets
		• Answering questions that require only yes, no, or a number as responses

210

Increased attention	Decreased attention	
• Reasoning in spatial contexts • Reasoning with proportions • Reasoning from graphs • Reasoning inductively and deductively	• Relying on outside authority (teacher or an answer key)	◄ **Reasoning**
• Connecting mathematics to other subjects and to the world outside the classroom • Connecting topics within mathematics • Applying mathematics	• Learning isolated topics • Developing skills out of context	◄ **Connections**
• Developing number sense • Developing operation sense • Creating algorithms and procedures • Using estimation both in solving problems and in checking the reasonableness of results • Exploring relationships among representations of, and operations on, whole numbers, fractions, decimals, integers, and rational numbers • Developing an understanding of ratio, proportion, and percent	• Memorizing rules and algorithms • Practicing tedious paper-and-pencil computations • Finding exact forms of answers • Memorizing procedures, such as cross-multiplication, without understanding • Practicing rounding numbers out of context	◄ **Number/Operations/ Computation**
• Identifying and using functional relationships • Developing and using tables, graphs, and rules to describe situations • Interpreting among different mathematical representations	• Topics seldom in the current curriculum	◄ **Patterns and Functions**

	Increased attention	**Decreased attention**
Algebra ▶	• Developing an understanding of variables, expressions, and equations • Using a variety of methods to solve linear equations and informally investigate inequalities and nonlinear equations	• Manipulating symbols • Memorizing procedures and drilling on equation solving
Statistics ▶	• Using statistical methods to describe, analyze, evaluate, and make decisions	• Memorizing formulas
Probability ▶	• Creating experimental and theoretical models of situations involving probabilities	• Memorizing formulas
Geometry ▶	• Developing an understanding of geometric objects and relationships • Using geometry in solving problems	• Memorizing geometric vocabulary • Memorizing facts and relationships
Measurement ▶	• Estimating and using measurement to solve problems	• Memorizing and manipulating formulas • Converting within and between measurement systems

212

Increased attention	Decreased attention	

- Actively involving students individually and in groups in exploring, conjecturing, analyzing, and applying mathematics in both a mathematics and real-world context

- Using appropriate technology for computation and exploration

- Using concrete materials

- Being a facilitator of learning

- Assessing learning as an integral part of instruction

- Teaching computations out of context

- Drilling on paper-and-pencil algorithms

- Teaching topics in isolation

- Stressing memorization

- Being the dispenser of knowledge

- Testing for the sole purpose of assigning grades

◀ **Instructional Practices**

Summary of Changes in Content and Emphasis for Mathematics in Grades Nine Through Twelve

- Use of real-world problems to motivate and apply theory

- Use of computer utilities to develop conceptual understanding

- Computer-based methods, such as successive approximations and graphing utilities, for solving equations and inequalities

- Structure of number systems

- Matrices and their applications

- Word problems by type, such as coin, digit, and work

- Simplification of radical expressions

- Use of factoring to solve equations and to simplify rational expressions

- Operations with rational expressions

- Paper-and-pencil graphing of equations by point plotting

- Logarithmic calculations, using tables and interpolation

- Solution of systems of equations, using determinants

- Conic sections

◀ **Algebra**

213

Increased attention	Decreased attention

Geometry ▶

Increased attention	Decreased attention
• Integration across topics at all grade levels	• Euclidean geometry as a complete axiomatic system
• Coordinate and transformation approaches	• Proofs of incidence and between-ness theorems
• Development of short sequences of theorems	• Geometry from a synthetic viewpoint
• Deductive arguments expressed orally and in sentence or paragraph form	• Two-column proofs
• Computer-based explorations of two-dimensional and three-dimensional figures	• Inscribed and circumscribed polygons
	• Theorems for circles involving segment ratios
• Three-dimensional geometry	• Analytic geometry as a separate course
• Realistic applications and modeling	

Trigonometry ▶

Increased attention	Decreased attention
• Use of appropriate scientific calculators	• Verification of complex identities
• Realistic applications and modeling	• Numerical applications of sum, difference, double-angle, and half-angle identities
• Connections among the right triangle ratios, trigonometric functions, and circular functions	• Calculations using tables and interpolation
• Use of graphing utilities for solving equations and inequalities	• Paper-and-pencil solutions of trigonometric equations

Functions ▶

Increased attention	Decreased attention
• Integration across topics at all grade levels	• Paper-and-pencil evaluation
• Connections among a problem situation, its model as a function in symbolic form, and the graph of that function	• Graphing of functions by hand, using tables of values
	• Formulas given as models of real-world problems
• Function equations expressed in standardized form as checks on the reasonableness of graphs produced by graphing utilities	• Expression of function equations in standardized form in order to graph them
• Functions constructed as models of real-world problems	• Treatment as a separate course

Increased attention	Decreased attention	
Yes		◀ **Statistics**
Yes		◀ **Probability**
Yes		◀ **Discrete Mathematics**

• Active involvement of students in constructing and applying mathematical ideas	• Teacher and text as exclusive sources of knowledge	◀ **Instructional Practices**
• Problem solving as a means as well as a goal of instruction	• Rote memorization of facts and procedures	
• Effective questioning techniques that promote student interaction	• Extended periods of individual seatwork practicing routine tasks	
• Use of a variety of instructional formats (small groups, individual explorations, peer instruction, whole-class discussions, project work)	• Instruction by teacher exposition	
	• Paper-and-pencil manipulative skill work	
• Use of calculators and computers as tools for learning and doing mathematics	• Relegation of testing to an adjunct role with the sole purpose of assigning grades	
• Student communication of mathematical ideas orally and in writing		
• Establishment and application of the interrelatedness of mathematical topics		
• Systematic maintenance of student learnings and embedding review in the context of new topics and problem situations		
• Assessment of learning as an integral part of instruction		

Selected References

Brutlag, D. *Beyond the Surface.* Investigations Project. Oakland: Office of the President, University of California, 1991.

A Call for Change: Recommendations for the Mathematical Preparation of Teachers of Mathematics. Washington, D.C.: Mathematical Association of America, 1990.

Caught in the Middle: Educational Reform for Young Adolescents in California Public Schools. Report of the Superintendent's Middle Grades Task Force. Sacramento: California Department of Education, 1987.

Cohen, E. G. *Designing Groupwork: Strategies for the Heterogeneous Classroom.* New York: Teachers College Press, 1986.

Crabill, C. *Cooperative Learning in Mathematics: A Handbook for Teachers.* Edited by N. Davidson. Menlo Park, Calif.: Addison Wesley, 1990.

Curriculum and Evaluation Standards for School Mathematics. Reston, Va.: National Council of Teachers of Mathematics, 1989.

Discrete Mathematics Across the Curriculum. Reston, Va.: National Council of Teachers of Mathematics, 1991.

"Do Bees Build It Best?" Interactive Mathematics Project. Berkeley: University of California; San Francisco: San Francisco State University, 1991.

Dossey, J. A., and others. *The Mathematics Report Card: Are We Measuring Up?* Princeton, N.J.: Educational Testing Service, 1988.

Educational Leadership, Vol. 46, No. 7 (April, 1989). Special issue on assessment.

Everybody Counts: A Report to the Nation on the Future of Mathematics Education. Washington, D.C.: National Academy Press, 1989.

Mathematics Assessment: Myths, Models, Good Questions, and Practical Suggestions. Edited by J. K. Stenmark. Reston, Va.: National Council of Teachers of Mathematics, Inc., 1991.

Mathematics Framework for California Public Schools, Kindergarten Through Grade Eight. Sacramento: California Department of Education, 1972.

Mathematics Framework for California Public Schools, Kindergarten Through Grade Twelve. Sacramento: California Department of Education, 1985.

Mathematics Model Curriculum Guide, Kindergarten Through Grade Eight. Sacramento: California Department of Education, 1987.

McKnight, C. C., and others. *The Underachieving Curriculum: Assessing U.S. School Mathematics from an International Perspective.* Edited by Ken Travers. Champaign, Ill.: Stipes Publishing Company, 1987. This report is sometimes referred to as "The SIMS Report"—a national report on the Second International Mathematics Study.

Mumme, J. *Portfolio Assessment in Mathematics.* Santa Barbara: Tri-County Mathematics Project, University of California, 1991.

Oakes, Jeannie. "Tracking: Can Schools Take a Different Route?" *NEA Today,* Vol. 6, No. 6 (January, 1988), 41–47.

On the Shoulders of Giants: New Approaches to Numeracy. Edited by L. A. Steen. Washington, D.C.: National Academy Press, 1990.

Professional Standards for Teaching Mathematics. Reston, Va.: National Council of Teachers of Mathematics, 1991.

Reshaping School Mathematics: A Philosophy and Framework for Curriculum. Washington, D.C.: National Academy Press, 1990.

Resnick, L. B. "Defining, Assessing, and Teaching Number Sense," in *Establishing Foundations for Research on Number Sense and Related Topics: Report of a Conference.* Edited by J. T. Sowder and B. P. Schappelle. San Diego: San Diego State University, 1989.

Resnick, L. B. *Education and Learning to Think.* Washington, D.C.: National Academy Press, 1987.

Russell, S. J., and others. *Beyond Drill and Practice: Expanding the Computer Mainstream.* Reston, Va.: The Council for Exceptional Children, 1989.

A Sampler of Mathematics Assessment. Sacramento: California Department of Education, 1991.

Statement on Competencies in Mathematics Expected of Entering Freshmen. Sacramento: California Department of Education, 1989.

Stenmark, J. K. *Assessment Alternatives in Mathematics.* Berkeley: Lawrence Hall of Science, 1989.

Stenmark, J. K., V. Thompson, and R. Cossey. *Family Math.* Berkeley: Lawrence Hall of Science, 1986.

Stigler, J. W., and H. W. Stevenson. "How Asian Teachers Polish Each Lesson to Perfection," *American Educator,* Vol. 15, No. 1 (Spring, 1991), 12–20, 43–47.

Thurston, W. "Mathematical Education," *Notices of the American Mathematical Society,* Vol. 37, No. 7 (September, 1990), 844–50.

Wiggins, G. "Teaching to the (Authentic) Test," *Educational Leadership,* Vol. 46, No. 7 (April, 1989), 41–47.

Mathematics Publications Available from the California Department of Education

The *Mathematics Framework* and other mathematics publications that complement it are available at cost from the California Department of Education.

ISBN	Title (Date of publication)	Price
0-8011-0777-6	The Changing Mathematics Curriculum: A Booklet for Parents (1989)*	$5.00/10
0-8011-0891-8	The Changing Mathematics Curriculum: A Booklet for Parents (Spanish Edition) (1991)*	5.00/10
0-8011-9974-0	Everybody Counts: A Report to the Nation on the Future of Mathematics Education (1989)	5.00
0-8011-9973-5	Everybody Counts: A Report to the Nation on the Future of Mathematics Education (1989) (2 to 9 copies)	4.25 per copy
0-8011-9972-7	Everybody Counts: A Report to the Nation on the Future of Mathematics Education (1989) (10 + copies)	3.00 per copy
0-8011-1033-5	Mathematics Framework for California Public Schools, Kindergarten Through Grade Twelve (1992)	6.75
0-8011-0664-8	Mathematics Model Curriculum Guide, Kindergarten Through Grade Eight (1987)	4.75
0-8011-0815-2	A Question of Thinking: A First Look at Students' Performance on Open-ended Questions in Mathematics (1989)	6.50
0-8011-0972-8	A Sampler of Mathematics Assessment (1991)	5.00
0-8011-0926-4	Seeing Fractions: Representations of Wholes and Parts (1990)	7.50
0-8011-0836-5	Statement on Competencies in Mathematics Expected of Entering Freshmen (1989)	5.00

*The price for 100 booklets is $30.00; the price for 1,000 booklets is $230.00.

Orders should be directed to:

Bureau of Publications
California Department of Education
P.O. Box 271
Sacramento, CA 95812-0271

Please include the International Standard Book Number (ISBN) for each title ordered.

Remittance or purchase order must accompany order. Purchase orders without checks are accepted only from governmental agencies. Sales tax should be added to all orders from California purchasers. Stated prices include shipping charges to anywhere in the United States.

A complete list of publications available from the Department may be obtained by writing to the address listed above or by calling (916) 445-1260.

93 84609   R93-34 (Third printing) 003-0064-93 300 10-93 45M